LETTERS
to a
YOUNG DOCTOR

By Richard Selzer

WITH A NEW PREFACE
BY THE AUTHOR

A HARVEST BOOK
HARCOURT BRACE & COMPANY
San Diego   New York   London

In memory of my father,
Julius Louis Selzer, M. D.

Requests for permission to make copies of any part of the work should
be mailed to: Permissions Department, Harcourt Brace & Company,
6277 Sea Harbor Drive, Orlando, Florida 32887-6777.

Parts of this book appeared in *The Atlantic,
The New England Review, The Iowa Review, Self,* and *Anteus.*

The author is grateful for permission to use an excerpt from
Homer, *The Odyssey*, translated by Robert Fitzgerald.
Copyright © 1961 by Robert Fitzgerald.
Reprinted by permission of Doubleday & Company, Inc.

Library of Congress Cataloging-in-Publication Data
Selzer, Richard, 1928–
Letters to a young doctor/by Richard Selzer.—1st Harvest ed.
p.   cm.—(A Harvest book)
Originally published: New York: Simon & Schuster, c1982
ISBN 0-15-600399-6
1. Surgery. 2. Medicine. 3. Residents (Medicine). I. Title.
RD39.S437   1996
617—dc20   95-53777

Designed by Christine Aulicono
Text set in Bulmer.

Printed in the United States of America
First Harvest edition
I K M O P N L J H

PREFACE
to the Harvest Edition (1996)

The title of this book was inspired by Rilke's *Letters to a Young Poet*. While these essays, memoirs, and stories are entirely lacking in the genius with which Rilke's *Letters* is infused, they were meant to be pedagogical and comradely—a reaching out to share. There was also the impulse of the singer, the troubadour who strikes his lyre and chants. Singing is the reason he is alive. More than anything else, these writings strive to tell what it is like to be a doctor, to tend the sick, to be sick.

Rilke was right when he wrote that the events of the body lie outside the precincts of language. Pain, for one, cannot be described, nor can the extreme of physical ecstasy. In order to approach these events, I have roamed the imagination for metaphor, myth, and memory. These pieces are not, strictly speaking, factual in the way that a medical textbook is factual. Nor are they journalism or reportage. They are literary renditions of medicine meant to strike resonating chords in the reader's mind. They are as much concerned with how something is said as they are with what is said. Style, here, is as important as content. Still, because I have spent a lifetime in medicine, there is much accurate detail. I have tried for anatomical and physiological fidelity but often without supplying a fixed meaning. I much prefer to invite interpretation.

I am grateful for the chance to see this work brought back into print. I cannot but see it as an act of affirmation. The real recognition has come from the generations of medical students and nurses who have read this book and written to tell me so. Not that there isn't a little melancholy in the sight of these many thousands of words that have fallen from my pen, filling sheet after sheet through multiple revisions such as to insist that I use paper the opposite side of which had been used before. (I never could justify the wastefulness that disdains half-used paper.) There is embarrassment as well. Writing may not be a crime, but it is always a lapse of taste. Mine anyway, which is explicit and shameless, a never-ending struggle between refinement and decadence, as one critic has commented.

This book was written in longhand, as is still my custom. It has to do with my inkwell, all that remains of an old Chinese inkstand. Black-and-red lacquered it was, and inlaid with mother-of-pearl. The lid of the inkwell is a bronze dragon that you pick up by the tail. Whenever you do, there is still the faint smell of sandalwood. A genie lives in this inkwell. Every time I fill my fountain pen, the scriptorium fills with aromatic smoke and he grants me three wishes. I usually ask for a metaphor or a bit of imagery. Unlike the genie in the *Arabian Nights,* mine is not a prisoner. He is enwelled in Paradise. Why not? All his wants are taken care of. Ink is for him nectar and ambrosia. His favorite brand is Higgins' Eternal, which had long gone out of manufacture until I found the recipe in an old book. Now I concoct it for him myself. And clean? He is the cleanest genie in the world, leaves only a bit of sludge at the bottom, which I aspirate with a medicine dropper once a year. He loves his little airings whenever I ink up. And he loves me too. I know because once, and only once, he broke from his usual reticence and told me that he would not wish to "remain intact" for one minute after my death. He would rather, he said, "diffuse." I was touched.

Each of these chapters is rooted in an event that I have experienced or witnessed. "Imelda" took place while on a surgical expedition to the Third World. The setting of "Toenails" was the New Haven Public Library, which I have been frequenting for forty years and where I still perform my humble acts of podiatry. "Mercy" was first remembered, then written at the Bellagio Study Center on Lake Como in Italy. The five "Letters to a Young Surgeon" are meant to be revelatory rather than instructive. Perhaps they serve to answer the question I have been asked so often: What is it like to lay open the body of a fellow human being? In "Textbook," "Rounds," and "Semiprivate, Female," I strove to photograph the essential patient and to nail that snapshot to the page. Some few of these pieces—the meditation on a slug, "The Grand Urinal of the Elks," and "A Pint of Blood"—were meant to entertain my students and colleagues. I take full blame for any failure to do just that. The writing of "Brute" was an act of atonement written a quarter of a century after my having abused a patient, an abuse for which I have not forgiven myself to this day. "Chatterbox" was inspired by a woman whose affliction was her need to talk all the time. It was written in homage to St. Catherine of Siena, whose letters to her confessor, Fra Raimondo, are among the masterpieces of religious mysticism. "Impostor" was my attempt to use myth and fable to lay bare the heart of a doctor. There is more than a hint of autobiography in it. In this, I have followed the advice of Petronius that an artist ought to use allusion, myth, and symbol so that his work will be like the prophecy of an inspired seer rather than the sworn statement made under oath before witnesses.

Throughout this volume there is a strong note of spirituality. It would be hard to find a page that is without it. This is only natural for a writer who sees the flesh as the spirit thickened. Within the damaged or fragile body where brute and painful facts have gathered, there is always the possibility that we will come to a new understanding and to perceive the body as a primal mystery and therefore sacred. Again and again, in patients deformed or ravaged by disease, we are stunned by a sudden radiance. This is not always comforting; there is terror in occasions that lift the veil from the ordinary world. Over my desk there is a message written by John Donne: " . . . our blood labors to beget spirits." I believe I know what he meant.

CONTENTS

CONTENTS

LETTERS
to a
YOUNG DOCTOR

Textbook

I send as your graduation present
my father's old textbook of physical diagnosis. It was published in
1918. Lifted yesterday from a trunk in the attic it is still faintly
redolent of formaldehyde, and stained with Heaven only knows
what ancient liquid. I love my old books—Longfellow, Virgil,
Romeo and Juliet and *Moby Dick*—but I love this *Textbook of
Physical Diagnosis* more. I can think of no better thing to give you
as a reminder that all of Medicine is a continuum of which you are
now a part. Within you is the gesture of the prehistoric surgeon who
trephined his neighbor's skull on the floor of a cave. Within you, the
poultice of cool mud applied to a burn by an old African woman.
The work of all doctors before you is in your blood. Yours will
enter the veins of whosoever comes after you.

The patients shown on the pages of this book are long since
dead; so, too, the doctors and nurses who tended them; so, too, the
photographer who peered at them from beneath the black curtain of

---

his camera. Gone, the brilliant pink chancre that gave to one man all of his distinction; gone, too, the great goiter that made of the most ordinary woman a grinning queen; and the boyish bravado that rose above a huge scrotum infested with microfilaria. To read this book is to understand that disease raises the sufferer, granting him from out of his fever and his fret an intimate vision of life, a more direct route to his soul. Now he has the body of a poet.

You cannot separate passion from pathology any more than you can separate a person's spirit from his body. Think of a particular person's spirit and, Presto! it is immediately incorporated. It has the size and shape of his body. The flesh is the spirit thickened. Gaze long enough at your patients, and from even the driest husk there will fly upward a shower of sparks that, to him who gazes, will coalesce into a little flame. Look at the pictures in this book and learn that the sick are refugees who must be treated kindly and gravely, without condescension. In the beginning you will love their wounds because they give you the occasion for virtue; later you will love the sick for their own sake. Rendered helpless by their afflictions, they cherish the memory of fertile lands and cool green glades and the company of love—all the stuff of their former selves. These people know something you and I do not yet know—what it is to live with the painful evidence of your mortality.

Notice that in each picture the eyes of the subjects have been covered by a black band to conceal identities. But the eyes are not the only windows to the soul. I have seen sorrow more fully expressed in a buttocks eaten away by bedsores; fear, in the arching of a neck; supplication, in a wrist. Only last week I was informed by a man's kneecaps that he was going to die. Flashing blue lights, they teletyped that he was running out of oxygen and blood. As soon as I got their cyanotic message, I summoned his family for a last vigil. A doctor's eyes must not be blindfolded against the light.

My hat is off to the photographer of this book, who chose misery for his subject in order to endear it. If I were you, I would not show these pictures to the squeamish who will be threatened by the echoes of their own mortality. Nor to vulgar people who decree that what they think ugly or gross ought not to be photographed. It would offend them. But what some think of as ugliness can become beauty to others—an ulcer, a dwarf, blood spreading on a pillow. An amputated leg retains something of the character of the one from whom it has been severed. Much as does the broken-off handle of a Greek amphora. It could have been part of nothing else. Retrieved and held up to fit, the handle sings again its old amphoric song. Personally, I suspect that truth is more accessible in "ugliness" than it is in beauty. The man who photographed the people in this book knew that in October, when the leaves fall from the trees, you can see farther into the forest.

Let us look through the book together. A quick glance might lead you to think these folk impoverished. My God! you say, a squirrel in his packed nuttery has more than they. Well, perhaps. But to me they seem the freest humans on earth. For while they stand on the narrow ledge of physical torment, no one would try to jostle them from their perch to take their place. Soon, you know, they will leap from that ledge to the plain where the bones of their ancestors lie strewn. The rest of us must still contend for a place in the sun, must face as best we can the dreadful future tense.

On one page there are two photographs of the same woman. She is naked save for her high-button shoes. In the first picture, she stands sideways to show her huge, pendulous umbilical hernia. It hangs to within six inches of her knees like an apron. In the second, she lifts the flap with her two hands as though to begin a dance. On the next page a man with a detachable collar and sleeve garters holds out a hand whose fingers are mummified from lack of circulation.

And here is a black man shown from the rear. His back is so wasted that he resembles a hanging bat—lax webs slung from the shoulder blades and cinched at the hips. His scapulae have scrambled to the top of the wreck, and crouch there in fright.

A young woman throws her head back to show her credential here, a thyroglossal duct cyst seen as a lump at the front of her neck. Her mouth falls open to reveal her small upper front teeth, each one separated by a cunning space. The interior of her mouth is dark, a receptacle. Her face is oddly corrupt. Perhaps she is kneeling . . . and those babyish sharp teeth.

Next, the reckless stare of a hyperthyroid, her eye showing too much white. She lives in a frenzy. Her pulse, respirations, appetite, everything is furious. Her gaze, too. Here, at last, is countenance befitting the mood of the body. Even the page trembles finely in my hand.

Now a congenital syphilitic intent that you should see his perforated palate. A triangular hole in that palate leads to an upper vault that is hidden from view, all but a bit of glistening membrane at one corner. It is a secret room at the very base of his brain. Of the man's face, you see only the nostrils and a Ubangi mouth stretched to accommodate the mirror inserted for the photography.

Here, the kyphotic, the scoliotic, the severely lordotic. Their faces are older by far than their S-shaped bodies, as though the curvature of a spine is associated with a compensatory prolongation of childhood. Childhood, it is said, ought to be prolonged by whatever means possible. But not at the cost of such crookedness.

And on this page, a youngish woman with cross-eyed breasts. A pouch of fatigue hangs beneath one nipple. The other breast is shrunken, turned inward in bewilderment. You cannot see her face or arms, which might allow you to draw some conclusions. But at her throat—a string of brave dark beads.

An old woman lies back in whorish recumbency, her knees flexed and having drifted apart. She wears white cotton stockings rolled below the knees and knotted there. Her hands reach down to open her vulva to show something beefy and red growing just inside. It has always been her role to be helpful.

All this famished flesh. Pale as a family of fatherless little boys. Saint Hildegard was right; God does not inhabit healthy bodies. Now, shut the book, close your eyes and hear the crimson thump of your heart.

If, in a darkened room, a doctor holds a bright light against a hollow part of the body, he can see through the outer tissues to the structures within—arteries, veins, projecting shelves of bone. In such a ruby gloom, he can distinguish between a hernia and a hydrocele of the testicle, or he can light up a sinus behind the brow to find the tumor there. Unlike surgery, which opens the body to direct examination, this transillumination gives an indirect vision, calling into play only the simplest perceptions of the doctor. The pictures in this book are a kind of transillumination. They hold a camera before the human body and capture through the covering layers the truth within.

Dear to me as it is, the *Textbook of Physical Diagnosis* is only a book, and cannot make of you a doctor.

In the matter of Physical Diagnosis greed is not a sin; it is a virtue. There can never be enough hearts and lungs to teach a Doctor his business. Do not rely upon the X-ray machine, the electrocardiograph or the laboratory to tell you what your hands, eyes and ears can find out, lest your senses atrophy from disuse. The machine does not exist that can take the place of the divining physician. The physical examination affords the opportunity to touch your patient. It gives the patient the opportunity to be touched by you. In this exchange, messages are sent from one to the other that, if your examination is performed with honesty and humility, will cause the divining powers of the Augurs to be passed on to you—their last heir.

Before long, you will lay your palm upon the back of a patient's chest, first on one side, then the other, and you will detect any diminution in the transmitted vibration of his voice. You will know, then, that in that place beneath your hand, the drawn breath does not fill a part of the lung that has become collapsed or carnified. Now listen there with your stethoscope. It is your cocked ear. Are there wheezes? bubbles? Do they occur as the breath is drawn or expelled? Are the bubbles coarse or fine? They will be fine only in the air sacs of the lung, coarse only in the large tubes—bronchi or trachea.

Again lay your palm upon the patient's chest such that your fingers contact him everywhere. Now tap your middle finger with the middle finger of your other hand. Listen . . . no, feel, it is something between listening and feeling that you do here—for the note that is struck. It ought to have a certain echo, a timbre. If it is not resonant, but instead the sound is flat and dead, something—fluid, a mass—is interposed between the lung and your hand. Tap out the area of dullness. Does it shift as he changes position? Why, then, like all fluid, it seeks its own level. Tap your way up the patient's back. At what place does the dullness change to resonance? The ninth rib? The eighth? Now you know how much fluid is present. You know, too, between which ribs your needle must pass in order to draw it out so that the compressed lung may expand. Even as you diagnose, you have begun your therapy.

Place your hand over the patient's heart. Let your fingers receive its beat. Find the exact place where the thrust is strongest. If this impulse is beyond the anterior fold of the armpit, the heart is enlarged. Should there be no single clean beat but a "thrill," as though a wren is stirring beneath your hand, the heart is damaged by disease or made wrong from the beginning. Now listen with the stethoscope. What you felt as a thrill is heard as a murmur. The wren sings. Touch and hearing blend, confirm. If the murmur be no soft whisper, but a harsh grate, or a rumble up and down the scale, a valve of the heart is damaged. Do you hear the slap of its calcified leaflets? Soon you will know which valve is tight, which leaky. Listening to the heart is like learning the songs of birds. A song once heard and identified is your own from then on. It will never be confused with the songs of other birds.

Is the rhythm crazy, coming willy-nilly, irregularly irregular? Then you have heard "fibrillation." Are the beats too close together to count? Perhaps this is "flutter." A bruit is a thrill that is heard. Here blood eddies around a narrows. Learn the language of this craft. Bruit, murmur, thrill, flutter, fibrillation—these are simple nouns that will soon become infused with knowledge and implication. Rejoice in these words that are used only in the highest purpose and that bind you to the others who do this work.

The abdomen is perhaps the deepest of mysteries. The very word *abdomen* has no known origin. It comes down to us from ancient times. Perhaps, they say, it means a secret place where things are stowed. You must approach such a mystery with tact. It is more threatening for a patient to uncover his abdomen than his chest. It is a kind of surrender. The abdomen, unprotected by a cage of ribs or a thick hide, presents itself equally to the surgeon's knife and the assassin's dagger. The slightest roughness on your part is a breach of the Articles of Medicine. In palpating the abdomen, the pressure of your hand must be neither too heavy nor too light. Too heavy a touch produces a guarding—a defense of the musculature. You will feel nothing through such a barricade. Too light, and you will tickle. This yields only discomfort and embarrassment. It goes without saying that your hands must be warm and dry. The belly is sensitive to every shock. Once intimidated, it can be reassured only at some effort, and then never fully. If the abdomen be divided into four quadrants, it is best to begin in the left upper quadrant, which is the least likely to be the site of disease, although by no means immune. Request that the patient roll himself a bit toward you. Guide him until he is at an angle of forty-five degrees. This allows the spleen to fall forward. Press just below the rib cage while asking him to take a deep breath. Such a full inspiration will depress the diaphragm and with it, the spleen. If the spleen is enlarged beyond normal size, its lower pole will bump against your fingers. How far below the ribs do you feel it? Two fingerbreadths? Three? From this, you may estimate its degree of enlargement. Beware! Many the spleen, outsized and therefore delicate, that has been ruptured by the undue pressure of the examining hand.

The most cerebral of all examinations is that of the brain. Encased as it is in a bony box, the brain cannot be approached directly, but its condition must be deduced from signals given at the most distant outposts. The navel, say, or the knee. Stroke the sole of a patient's foot. Stroke it firmly with a hard instrument, and see the toes of that foot rise and spread. This is the sign of Babinski. It tells, perhaps, of a blood clot or tumor in the motor cortex of the opposite side of the brain. Ask your patient to stand and close his eyes. "I won't let you fall," you assure him. Almost at once he sways to one side and would surely fall if you did not catch him. So! the cerebellum too is involved. You are closing in on the lesion.

Soon you will know its exact location. Now you are happy in the way that a hunter is happy when, following the spoor of an animal, he comes upon fresh evidence.

Just so, in the name of Asklepios, do I invite you to begin the sacred process of divination called Physical Diagnosis. There is no more beautiful sight in the world than that of a kindly, efficient doctor engaged in the examination of the body of a fellow human being.

Imelda

———————•◦•———————

I heard the other day that Hugh
Franciscus had died. I knew him once. He was the Chief of Plastic
Surgery when I was a medical student at Albany Medical College.
Dr. Franciscus was the archetype of the professor of surgery—tall,
vigorous, muscular, as precise in his technique as he was impec-
cable in his dress. Each day a clean lab coat monkishly starched,
that sort of thing. I doubt that he ever read books. One book only,
that of the human body, took the place of all others. He never
raised his eyes from it. He read it like a printed page as though he
knew that in the calligraphy there just beneath the skin were all the
secrets of the world. Long before it became visible to anyone else,
he could detect the first sign of granulation at the base of a wound,
the first blue line of new epithelium at the periphery that would tell
him that a wound would heal, or the barest hint of necrosis that
presaged failure. This gave him the appearance of a prophet. "This

skin graft will take," he would say, and you must believe beyond all cyanosis, exudation and inflammation that it would.

He had enemies, of course, who said he was arrogant, that he exalted activity for its own sake. Perhaps. But perhaps it was no more than the honesty of one who knows his own worth. Just look at a scalpel, after all. What a feeling of sovereignty, megalomania even, when you know that it is you and you alone who will make certain use of it. It was said, too, that he was a ladies' man. I don't know about that. It was all rumor. Besides, I think he had other things in mind than mere living. Hugh Franciscus was a zealous hunter. Every fall during the season he drove upstate to hunt deer. There was a glass-front case in his office where he showed his guns. How could he shoot a deer? we asked. But he knew better. To us medical students he was someone heroic, someone made up of several gods, beheld at a distance, and always from a lesser height. If he had grown accustomed to his miracles, we had not. He had no close friends on the staff. There was something a little sad in that. As though once long ago he had been flayed by friendship and now the slightest breeze would hurt. Confidences resulted in dishonor. Perhaps the person in whom one confided would scorn him, betray. Even though he spent his days among those less fortunate, weaker than he—the sick, after all—Franciscus seemed aware of an air of personal harshness in his environment to which he reacted by keeping his own counsel, by a certain remoteness. It was what gave him the appearance of being haughty. With the patients he was forthright. All the facts laid out, every question anticipated and answered with specific information. He delivered good news and bad with the same dispassion.

I was a third-year student, just turned onto the wards for the first time, and clerking on Surgery. Everything—the operating room, the morgue, the emergency room, the patients, professors, even the nurses—was terrifying. One picked one's way among the mines and booby traps of the hospital, hoping only to avoid the hemorrhage and perforation of disgrace. The opportunity for humiliation was everywhere.

It all began on Ward Rounds. Dr. Franciscus was demonstrating a cross-leg flap graft he had constructed to cover a large fleshy defect in the leg of a merchant seaman who had injured himself in a fall. The man was from Spain and spoke no English. There had been a comminuted fracture of the femur, much soft tissue damage, necrosis. After weeks of débridement and dressings, the wound had been made ready for grafting. Now the patient was in his fifth postoperative day. What we saw was a thick web of pale blue flesh arising from the man's left thigh, and which had been sutured to the open wound on the right thigh. When the surgeon pressed the pedicle with his finger, it blanched; when he let up, there was a slow return of the violaceous color.

"The circulation is good," Franciscus announced. "It will get better." In several weeks, we were told, he would divide the tube of flesh at its site of origin, and tailor it to fit the defect to which, by then, it would have grown more solidly. All at once, the webbed man in the bed reached out, and gripping Franciscus by the arm, began to speak rapidly, pointing to his groin and hip. Franciscus stepped back at once to disengage his arm from the patient's grasp.

"Anyone here know Spanish? I didn't get a word of that."

"The cast is digging into him up above," I said. "The edges of the plaster are rough. When he moves, they hurt."

Without acknowledging my assistance, Dr. Franciscus took a plaster shears from the dressing cart and with several large snips cut away the rough edges of the cast.

"*Gracias, gracias.*" The man in the bed smiled. But Franciscus had already moved on to the next bed. He seemed to me a man of immense strength and ability, yet without affection for the patients. He did not want to be touched by them. It was less kindness that he showed them than a reassurance that he would never give up, that he would bend every effort. If anyone could, he would solve the problems of their flesh.

Ward Rounds had disbanded and I was halfway down the corridor when I heard Dr. Franciscus' voice behind me.

"You speak Spanish." It seemed a command.

"I lived in Spain for two years," I told him.

"I'm taking a surgical team to Honduras next week to operate on the natives down there. I do it every year for three weeks, some-where. This year, Honduras. I can arrange the time away from your duties here if you'd like to come along. You will act as interpreter. I'll show you how to use the clinical camera. What you'd see would make it worthwhile."

So it was that, a week later, the envy of my classmates, I joined the mobile surgical unit—surgeons, anesthetists, nurses and equipment—aboard a Military Air Transport plane to spend three weeks performing plastic surgery on people who had been previ-ously selected by an advance team. Honduras. I don't suppose I shall ever see it again. Nor do I especially want to. From the plane it seemed a country made of clay—burnt umber, raw sienna, dry. It had a deadweight quality, as though the ground had no buoyancy, no air sacs through which a breeze might wander. Our destination was Comayagua, a town in the Central Highlands. The town itself was situated on the edge of one of the flatlands that were linked in a network between the granite mountains. Above, all was brown, with only an occasional Spanish cedar tree; below, patches of luxuriant tropical growth. It was a day's bus ride from the airport. For hours, the town kept appearing and disappearing with the convolutions of the road. At last, there it lay before us, panting and exhausted at the bottom of the mountain.

That was all I was to see of the countryside. From then on, there was only the derelict hospital of Comayagua, with the smell of spoiling bananas and the accumulated odors of everyone who had been sick there for the last hundred years. Of the two, I much preferred the frank smell of the sick. The heat of the place was incendiary. So hot that, as we stepped from the bus, our own words did not carry through the air, but hung limply at our lips and chins. Just in front of the hospital was a thirsty courtyard where mobs of waiting people squatted or lay in the meager shade, and where, on dry days, a fine dust rose through which untethered goats shoul-dered. Against the walls of this courtyard, gaunt, dejected men stood, their faces, like their country, preternaturally solemn, leaden. Here no one looked up at the sky. Every head was bent beneath a wide-brimmed straw hat. In the days that followed, from the doorway of the dispensary, I would watch the brown mountains sliding about, drinking the hospital into their shadow as the afternoon grew later and later, flattening us by their very altitude.

The people were mestizos, of mixed Spanish and Indian blood. They had flat, broad, dumb museum feet. At first they seemed to me indistinguishable the one from the other, without animation. All the vitality, the hidden sexuality, was in their black hair. Soon I was to know them by the fissures with which each face was graven. But, even so, compared to us, they were masked, shut away. My job was to follow Dr. Franciscus around, photograph the patients before and after surgery, interpret and generally act as aide-de-camp. It was exhilarating. Within days I had decided that I was not just useful, but essential. Despite that we spent all day in each other's company, there were no overtures of friendship from Dr. Franciscus. He knew my place, and I knew it, too. In the afternoon he examined the patients scheduled for the next day's surgery. I would call out a name from the doorway to the examining room. In the courtyard someone would rise. I would usher the patient in, and nudge him to the examining table where Franciscus stood, always, I thought, on the verge of irritability. I would read aloud the case history, then wait while he carried out his examination. While I took the "before" photographs, Dr. Franciscus would dictate into a tape recorder:

"Ulcerating basal cell carcinoma of the right orbit—six by eight centimeters—involving the right eye and extending into the floor of the orbit. Operative plan: wide excision with enucleation of the eye. Later, bone and skin grafting." The next morning we would be in the operating room where the procedure would be carried out.

We were more than two weeks into our tour of duty—a few days to go—when it happened. Earlier in the day I had caught sight of her through the window of the dispensary. A thin, dark Indian girl about fourteen years old. A figurine, orange-brown, terra-cotta, and still attached to the unshaped clay from which she had been carved. An older, sun-weathered woman stood behind and somewhat to the left of the girl. The mother was short and dumpy. She wore a broad-brimmed hat with a high crown, and a shapeless dress like a cassock. The girl had long, loose black hair. There were tiny gold hoops in her ears. The dress she wore could have been her mother's. Far too big, it hung from her thin shoulders at some risk of slipping down her arms. Even with her in it, the dress was empty, something hanging on the back of a door. Her breasts made only the smallest imprint in the cloth, her hips none at all. All the while, she pressed to her mouth a filthy, pink, balled-up rag as though to stanch a flow or buttress against pain. I knew that what she had come to show us, what we were there to see, was hidden beneath that pink cloth. As I watched, the woman handed down to her a gourd from which the girl drank, lapping like a dog. She was the last patient of the day. They had been waiting in the courtyard for hours.

"Imelda Valdez," I called out. Slowly she rose to her feet, the cloth never leaving her mouth, and followed her mother to the examining-room door. I shooed them in.

"You sit up there on the table," I told her. "Mother, you stand over there, please." I read from the chart:

"This is a fourteen-year-old girl with a complete, unilateral, left-sided cleft lip and cleft palate. No other diseases or congenital defects. Laboratory tests, chest X ray—negative."

"Tell her to take the rag away," said Dr. Franciscus. I did, and the girl shrank back, pressing the cloth all the more firmly.

"Listen, this is silly," said Franciscus. "Tell her I've got to see it. Either she behaves, or send her away."

"Please give me the cloth," I said to the girl as gently as possible. She did not. She could not. Just then, Franciscus reached up and, taking the hand that held the rag, pulled it away with a hard jerk. For an instant the girl's head followed the cloth as it left her face, one arm still upflung against showing. Against all hope, she would hide herself. A moment later, she relaxed and sat still. She seemed to me then like an animal that looks outward at the infinite, at death, without fear, with recognition only.

Set as it was in the center of the girl's face, the defect was utterly hideous—a nude rubbery insect that had fastened there. The upper lip was widely split all the way to the nose. One white tooth perched upon the protruding upper jaw projected through the hole. Some of the bone seemed to have been gnawed away as well. Above the thing, clear almond eyes and long black hair reflected the light. Below, a slender neck where the pulse trilled visibly. Under our gaze the girl's eyes fell to her lap where her hands lay palms upward, half open. She was a beautiful bird with a crushed beak. And tense with the expectation of more shame.

"Open your mouth," said the surgeon. I translated. She did so, and the surgeon tipped back her head to see inside.

"The palate, too. Complete," he said. There was a long silence. At last he spoke.

"What is your name?" The margins of the wound melted until she herself was being sucked into it.

"Imelda." The syllables leaked through the hole with a slosh and a whistle.

"Tomorrow," said the surgeon, "I will fix your lip. *Mañana.*"

It seemed to me that Hugh Franciscus, in spite of his years of experience, in spite of all the dreadful things he had seen, must have been awed by the sight of this girl. I could see it flit across his face for an instant. Perhaps it was her small act of concealment, that he had had to demand that she show him the lip, that he had had to force her to show it to him. Perhaps it was her resistance that intensified the disfigurement. Had she brought her mouth to him willingly, without shame, she would have been for him neither more nor less than any other patient.

He measured the defect with calipers, studied it from different angles, turning her head with a finger at her chin.

"How can it ever be put back together?" I asked.

"Take her picture," he said. And to her, "Look straight ahead." Through the eye of the camera she seemed more pitiful than ever, her humiliation more complete.

"Wait!" The surgeon stopped me. I lowered the camera. A strand of her hair had fallen across her face and found its way to her mouth, becoming stuck there by saliva. He removed the hair and secured it behind her ear.

"Go ahead," he ordered. There was the click of the camera. The girl winced.

"Take three more, just in case."

When the girl and her mother had left, he took paper and pen and with a few lines drew a remarkable likeness of the girl's face.

"Look," he said. "If this dot is A, and this one B, this, C and this, D, the incisions are made A to B, then C to D. CD must equal AB. It is all equilateral triangles." All well and good, but then came X and Y and rotation flaps and the rest.

"Do you see?" he asked.

"It is confusing," I told him.

"It is simply a matter of dropping the upper lip into a normal position, then crossing the gap with two triangular flaps. It is geometry," he said.

"Yes," I said. "Geometry." And relinquished all hope of becoming a plastic surgeon.

In the operating room the next morning the anesthesia had already been administered when we arrived from Ward Rounds. The tube emerging from the girl's mouth was pressed against her lower lip to be kept out of the field of surgery. Already, a nurse was scrubbing the face which swam in a reddish-brown lather. The tiny gold earrings were included in the scrub. Now and then, one of them gave a brave flash. The face was washed for the last time, and dried. Green towels were placed over the face to hide everything but the mouth and nose. The drapes were applied.

"Calipers!" The surgeon measured, locating the peak of the distorted Cupid's bow.

"Marking pen!" He placed the first blue dot at the apex of the bow. The nasal sills were dotted; next, the inferior philtral dimple, the vermilion line. The A flap and the B flap were outlined. On he worked, peppering the lip and nose, making sense out of chaos, realizing the lip that lay waiting in that deep essential pink, that only he could see. The last dot and line were placed. He was ready.

"Scalpel!" He held the knife above the girl's mouth.

"O.K. to go ahead?" he asked the anesthetist.

"Yes."

He lowered the knife.

"No! Wait!" The anesthetist's voice was tense, staccato. "Hold it!"

The surgeon's hand was motionless.

"What's the matter?"

"Something's wrong. I'm not sure. God, she's hot as a pistol. Blood pressure is way up. Pulse one eighty. Get a rectal temperature." A nurse fumbled beneath the drapes. We waited. The nurse retrieved the thermometer.

"One hundred seven . . . no . . . eight." There was disbelief in her voice.

"Malignant hyperthermia," said the anesthetist. "Ice! Ice! Get lots of ice!" I raced out the door, accosted the first nurse I saw.

"Ice!" I shouted. "*Hielo!* Quickly! *Hielo!*" The woman's expression was blank. I ran to another. "*Hielo! Hielo!* For the love of God, ice."

"*Hielo?*" She shrugged. "*Nada.*" I ran back to the operating room.

"There isn't any ice," I reported. Dr. Franciscus had ripped off his rubber gloves and was feeling the skin of the girl's abdomen. Above the mask his eyes were the eyes of a horse in battle.

"The EKG is wild . . ."

"I can't get a pulse . . ."

"What the hell . . ."

The surgeon reached for the girl's groin. No femoral pulse.

"EKG flat. My God! She's dead!"

"She can't be."

"She is."

The surgeon's fingers pressed the groin where there was no

pulse to be felt, only his own pulse hammering at the girl's flesh to be let in.

It was noon, four hours later, when we left the operating room. It was a day so hot and humid I felt steamed open like an envelope. The woman was sitting on a bench in the courtyard in her dress like a cassock. In one hand she held the piece of cloth the girl had used to conceal her mouth. As we watched, she folded it once neatly, and then again, smoothing it, cleaning the cloth which might have been the head of the girl in her lap that she stroked and consoled.

"I'll do the talking here," he said. He would tell her himself, in whatever Spanish he could find. Only if she did not understand was I to speak for him. I watched him brace himself, set his shoulders. How could he tell her? I wondered. What? But I knew he would tell her everything, exactly as it had happened. As much for himself as for her, he needed to explain. But suppose she screamed, fell to the ground, attacked him, even? All that hope of love . . . gone. Even in his discomfort I knew that he was teaching me. The way to do it was professionally. Now he was standing above her. When the woman saw that he did not speak, she lifted her eyes and saw what he held crammed in his mouth to tell her. She knew, and rose to her feet.

"*Señora,*" he began, "I am sorry." All at once he seemed to me shorter than he was, scarcely taller than she. There was a place at the crown of his head where the hair had grown thin. His lips were stones. He could hardly move them. The voice dry, dusty.

"No one could have known. Some bad reaction to the medicine for sleeping. It poisoned her. High fever. She did not wake up." The last, a whisper. The woman studied his lips as though she were deaf. He tried, but could not control a twitching at the corner of his mouth. He raised a thumb and forefinger to press something back into his eyes.

"*Muerte,*" the woman announced to herself. Her eyes were human, deadly.

"*Sí, muerte.*" At that moment he was like someone cast, still alive, as an effigy for his own tomb. He closed his eyes. Nor did he open them until he felt the touch of the woman's hand on his arm, a touch from which he did not withdraw. Then he looked and saw the grief corroding her face, breaking it down, melting the features so that eyes, nose, mouth ran together in a distortion, like the girl's. For a long time they stood in silence. It seemed to me that minutes passed. At last her face cleared, the features rearranged themselves. She spoke, the words coming slowly to make certain that he understood her. She would go home now. The next day her sons would come for the girl, to take her home for burial. The doctor must not be sad. God has decided. And she was happy now that the harelip had been fixed so that her daughter might go to Heaven without it. Her bare feet retreating were the felted pads of a great bereft animal.

The next morning I did not go to the wards, but stood at the gate leading from the courtyard to the road outside. Two young men in striped ponchos lifted the girl's body wrapped in a straw mat onto the back of a wooden cart. A donkey waited. I had been drawn to this place as one is drawn, inexplicably, to certain scenes of desolation—executions, battlefields. All at once, the woman looked up and saw me. She had taken off her hat. The heavy-hanging coil of her hair made her head seem larger, darker, noble. I pressed some money into her hand.

"For flowers," I said. "A priest." Her cheeks shook as though minutes ago a stone had been dropped into her navel and the ripples were just now reaching her head. I regretted having come to that place.

"*Sí, sí,*" the woman said. Her own face was stitched with flies. "The doctor is one of the angels. He has finished the work of God. My daughter is beautiful."

What could she mean! The lip had not been fixed. The girl had died before he would have done it.

"Only a fine line that God will erase in time," she said.

I reached into the cart and lifted a corner of the mat in which the girl had been rolled. Where the cleft had been there was now a fresh line of tiny sutures. The Cupid's bow was delicately shaped, the vermilion border aligned. The flattened nostril had now the same rounded shape as the other one. I let the mat fall over the face of the dead girl, but not before I had seen the touching place where the finest black hairs sprang from the temple.

"*Adiós, adiós* . . ." And the cart creaked away to the sound of hooves, a tinkling bell.

There are events in a doctor's life that seem to mark the boundary between youth and age, seeing and perceiving. Like certain dreams, they illuminate a whole lifetime of past behavior. After such an event, a doctor is not the same as he was before. It had seemed to me then to have been the act of someone demented, or at least insanely arrogant. An attempt to reorder events. Her death had come to him out of order. It should have come after the lip had been repaired, not before. He could have told the mother that, no, the lip had not been fixed. But he did not. He said nothing. It had been an act of omission, one of those strange lapses to which all of us are subject and which we live to regret. It must have been then, at that moment, that the knowledge of what he would do appeared to him. The words of the mother had not consoled him; they had hunted him down. He had not done it for her. The dire necessity was his. He would not accept that Imelda had died before he could repair her lip. People who do such things break free from society. They follow their own lonely path. They have a secret which they can never reveal. I must never let on that I knew.

How often I have imagined it. Ten o'clock at night. The hospital of Comayagua is all but dark. Here and there lanterns tilt and skitter up and down the corridors. One of these lamps breaks free from the others and descends the stone steps to the underground room that is the morgue of the hospital. This room wears the expression as if it had waited all night for someone to come. No silence so deep as this place with its cargo of newly dead. Only the slow drip of water over stone. The door closes gassily and clicks shut. The lock is turned. There are four tables, each with a body encased in a paper shroud. There is no mistaking her. She is the smallest. The surgeon takes a knife from his pocket and slits open the paper shroud, that part in which the girl's head is enclosed. The wound seems to be living on long after she has died. Waves of heat emanate from it, blurring his vision. All at once, he turns to peer over his shoulder. He sees nothing, only a wooden crucifix on the wall.

He removes a package of instruments from a satchel and arranges them on a tray. Scalpel, scissors, forceps, needle holder. Sutures and gauze sponges are produced. Stealthy, hunched, engaged, he begins. The dots of blue dye are still there upon her mouth. He raises the scalpel, pauses. A second glance into the darkness. From the wall a small lizard watches and accepts. The first cut is made. A sluggish flow of dark blood appears. He wipes it away with a sponge. No new blood comes to take its place. Again and again he cuts, connecting each of the blue dots until the whole of the zigzag slice is made, first on one side of the cleft, then on the other. Now the edges of the cleft are lined with fresh tissue. He sets down the scalpel and takes up scissors and forceps, undermining the little flaps until each triangle is attached only at one side. He rotates each flap into its new position. He must be certain that they can be swung without tension. They can. He is ready to suture. He fits the tiny curved needle into the jaws of the needle holder. Each suture is placed precisely the same number of millimeters from the cut edge, and the same distance apart. He ties each knot down until the edges are apposed. Not too tightly. These are the most meticulous sutures of his life. He cuts each thread close to the knot. It goes well. The vermilion border with its white skin roll is exactly aligned. One more stitch and the Cupid's bow appears as if by magic. The man's face shines with moisture. Now the nostril is incised around the margin, released, and sutured into a round shape to match its mate. He wipes the blood from the face of the girl with gauze that he has dipped in water. Crumbs of light are scattered on the girl's face. The shroud is folded once more about her. The instruments are handed into the satchel. In a moment the morgue is dark and a lone lantern ascends the stairs and is extinguished.

Six weeks later I was in the darkened amphitheater of the Medical School. Tiers of seats rose in a semicircle above the small stage where Hugh Franciscus stood presenting the case material he had encountered in Honduras. It was the highlight of the year. The hall was filled. The night before he had arranged the slides in the order in which they were to be shown. I was at the controls of the slide projector.

"Next slide!" he would order from time to time in that military voice which had called forth blind obedience from generations of medical students, interns, residents and patients.

"This is a fifty-seven-year-old man with a severe burn contracture of the neck. You will notice the rigid webbing that has fused the chin to the presternal tissues. No motion of the head on the torso is possible. . . . Next slide!"

"Click," went the projector.

"Here he is after the excision of the scar tissue and with the head in full extension for the first time. The defect was then covered. . . . Next slide!"

"Click."

". . . with full-thickness drums of skin taken from the abdomen with the Padgett dermatome. Next slide!"

"Click."

And suddenly there she was, extracted from the shadows, suspended above and beyond all of us like a resurrection. There was the oval face, the long black hair unbraided, the tiny gold hoops in her ears. And that luminous gnawed mouth. The whole of her life seemed to have been summed up in this photograph. A long silence followed that was the surgeon's alone to break. Almost at once, like the anesthetist in the operating room in Comayagua, I knew that something was wrong. It was not that the man would not speak as that he could not. The audience of doctors, nurses and students seemed to have been infected by the black, limitless silence. My own pulse doubled. It was hard to breathe. Why did he not call out for the next slide? Why did he not save himself? Why had he not removed this slide from the ones to be shown? All at once I knew that he had used his camera on her again. I could see the long black shadows of her hair flowing into the darker shadows of the morgue. The sudden blinding flash . . . The next slide would be the one taken in the morgue. He would be exposed.

In the dim light reflected from the slide, I saw him gazing up at her, seeing not the colored photograph, I thought, but the negative of it where the ghost of the girl was. For me, the amphitheater had become Honduras. I saw again that courtyard littered with patients. I could see the dust in the beam of light from the projector. It was then that I knew that she was his measure of perfection and pain— the one lost, the other gained. He, too, had heard the click of the camera, had seen her wince and felt his mercy enlarge. At last he spoke.

"Imelda." It was the one word he had heard her say. At the sound of his voice I removed the next slide from the projector. "Click" . . . and she was gone. "Click" again, and in her place the man with the orbital cancer. For a long moment Franciscus looked up in my direction, on his face an expression that I have given up trying to interpret. Gratitude? Sorrow? It made me think of the gaze of the girl when at last she understood that she must hand over to him the evidence of her body.

"This is a sixty-two-year-old man with a basal cell carcinoma of the temple eroding into the bony orbit . . ." he began as though nothing had happened.

At the end of the hour, even before the lights went on, there was loud applause. I hurried to find him among the departing crowd. I could not. Some weeks went by before I caught sight of him. He seemed vaguely convalescent, as though a fever had taken its toll before burning out.

Hugh Franciscus continued to teach for fifteen years, although he operated a good deal less, then gave it up entirely. It was as though he had grown tired of blood, of always having to be involved with blood, of having to draw it, spill it, wipe it away, stanch it. He was a quieter, softer man, I heard, the ferocity diminished. There were no more expeditions to Honduras or anywhere else.

I, too, have not been entirely free of her. Now and then, in the years that have passed, I see that donkey-cart cortège, or his face bent over hers in the morgue. I would like to have told him what I now know, that his unrealistic act was one of goodness, one of those small, persevering acts done, perhaps, to ward off madness. Like lighting a lamp, boiling water for tea, washing a shirt. But, of course, it's too late now.

Letter to a Young Surgeon

I

So. You have chosen Surgery.
Have you thought long and hard upon it? How necessary is the practice of Surgery to you? Would you die if you were not to do it? If you perceive Surgery as the loftiest branch of Medicine, remember that it is the one most vulnerable to injury and ignominy. It is not the privet hedge that is uprooted in a hurricane; it is the royal palm.

If all Medicine be considered a religion, why, then, the psychiatrist is the cloistered nun, a contemplative whose pale hands are unused save for the telling of beads. The surgeon is burly, brawling Friar Tuck, out in the world, taking up his full share of space, and always at some risk of exposing rather too much of his free-swinging beef from beneath his habit.

As for myself, it was never for a moment given me to choose my vocation. I could hardly have engaged myself in the Future, since a converse with Physics and Mathematics is essential when plotting the trajectory of a spacecraft. From both of these mysteries I have been forbidden by the gods. In hands such as mine, orbits are likely to become obits. In Physics I claim for my own only Archimedes' principle, having over and again witnessed its reenactment in my house. A child of mine, having filled the bathtub to the rim absolute, leaps in, thereby displacing to the bathroom floor an amount of water which I do not dispute is equal to the volume of the creature.

As for immersion in the Past, History or Literature? No. I am ill set up for the studies, since I am incapable of distinguishing among the preceding three centuries. The seventeenth, eighteenth and nineteenth are to me as one. All of their wars, pestilences, famines, artists and writers wallow together in my unselective memory. Did Bacon precede Lamb? Or the other way around? There is a vegetarian of my acquaintance who forbears to read either one. As for the present, it is all a matter of Psychology, and I could not bear to listen to the secret urges of others when my own so outdo them in flamboyance. I have always preferred what people do alone or in the company of one like-minded companion to the shifts and rumble of the mob. The only thing left, you see, was Surgery. But there is another reason.

One spring, when my brother Billy was ten years old, and I eight, the Hudson River at Troy, New York, turned traitor and overflowed its banks. Fifth Avenue between Jacob and Federal was inundated. Ecstasy! Children understand floods. To a child it is proper and amusing that the arrogant pavement, the high-and-mighty bricks and mortar and the malignant automobile should be humbled by the unexpected slap of a big water.

One morning Billy and I leaned out of the second-story bay window and looked out over a vast, gleaming sea. All at once we spied a rowboat rounding the corner from Federal Street. Up the block it beetled. A lone man pulled at the oars for all he was worth. Oh, how he rowed! Opposite our house, the man turned the rowboat in our direction and headed straight for the place where once our front stoop had been, and where now there raced a cold and filthy current. There was a confusion of bumps and scrapes, the sound of water splashing, the voices of men shouting. The next thing we saw was the little rowboat retreating from the house and making its way down the street. But now there were two men in it—the one who had come and . . . Father, perched in the stern. Wearing his gray fedora and long overcoat, he held his black medical bag on his knees. Father was making a house call in a rowboat! It was such a sight as the children of astronauts must have as their daily breakfast, lunch and dinner.

What, we were dying to know, was he going to do? Deliver a baby? Set a broken leg? Take out an appendix?

"I bet it's an appendix," said Billy.

"I bet it's a broken leg," I said.

"Nah, an appendix," reiterated Billy. "Then he could use it for bait."

"Yeah," I said, hooked as usual by the metaphor. "I bet it's an appendix, too."

I do not remember, if ever I was told, what it was that Father did on that voyage. But I shall never forget the sight of him sailing down Fifth Avenue toward Federal Street to save a life. It was my first real proof that Father was a hero.

Father died five years later. I was thirteen.

In the interview prior to one's acceptance into medical school, one is asked the obligatory question: Why do you want to become a doctor? I don't know what you said, but my answer had to do with Virgil. In the *Aeneid*, there is an old doctor who arrives at the siege of Troy to tend the wound of Aeneas, who has been struck by an arrow. The doctor's name is Iapyx. Now it happened that when Iapyx was a boy, Apollo fell in love with him, and offered him, as a gift, music, wisdom, prophecy or swift arrows. Iapyx chose none of these, and asked for Medicine instead. For he wished only to prolong the life of the father he loved. But as the battle for Troy raged all around him, Iapyx realized that he could not save Aeneas. All his skill was to no avail. He cried out to the gods to help him. Suddenly, the arrow, of its own accord, fell from the wound of

Aeneas. Iapyx surmised that more than man had wrought this cure. Iapyx was right. Venus, the mother of Aeneas, had placed a healing herb in the water that Iapyx was using to bathe the wound of her son, and the arrow was miraculously extruded.

I became a doctor to prolong my father's life, and many times since, I have summoned the gods to my side for consultation.

You may as well be told right now that Surgery as a healing art is a passing phenomenon. It may already have seen its time of greatest glory. I should not be surprised to learn that news of this waning might give rise to some melancholy in one who is just now embarking on a career as a surgeon, for he will surely outlive his usefulness, to become master of a dead art.

But what a joy this same news will be to the rest of mankind, to him who harbors a stone in his gallbladder and needs not to be cut for it, to her who has a cancer in her breast and needs not come to mastectomy. To them it shall be a blessing for which even the greediest, least altruistic surgeon must pray. Even now, the surgery that removes lung tissue infected with tuberculosis is largely a thing of the past. Operations for thyroid disease and peptic ulcer, which only two decades ago comprised the bulk of any operating-room schedule, are rarities. The internist, cardiologist and endocrinologist have grown more and more adept at controlling disease with medication, diet and prevention. Soon we surgeons will be turned out to pasture, to graze out our days dreaming of old wars and trying to remember what exactly they were all about. Only in Plastic and Reconstructive Surgery and the transplantation of organs will remnants of surgery survive. *Grace à Dieu,* I shall be safe in my tomb like an ancient king in moldered ermine, unmindful that the kingdom I once ruled had vanished long ago.

What is the difference between a surgeon and an internist? The surgeon, armed to the teeth, seeks to overwhelm and control the body; the medical man strives with pills and potions to cooperate with that body, even to the point of making concessions to disease.

One is the stance of the warrior; the other, that of the statesman. The more technologically oriented the internist becomes, the more like a surgeon he becomes, removing himself from the quiet resonances of the consulting room and bedside in order to flail away with his own machete. Heaven help the internist. He is not suited for such red labor. He is not blooded to it. And Heaven help his patient who, instead of the touch and gaze of wisdom, will receive an endoscope in his throat or a catheter in his heart.

Tomorrow you are an intern. You will work both night and day. You will be tired, I know, and I don't want any more of that kind of tired. An intern, like a poet, is at the disposal of his night. The more one drinks of the night, the more one thirsts for the light of day. This is as true for the doctor as it is for the insomniac. Still, there is something to be said for working at night in a hospital. The external glare of daytime is gone, and one is permeated by a wave of darkness. Now the ritual of healing is more naturally practiced. For so many years I have been patrolling these night corridors with the lamplight creeping just ahead from room to room, illuminating wounds that are faces, faces that are wounds. As though I were a sentry pacing out the border between home and a wild country. You must remember when you are tired and it is late at night, when the patients ask whether you are a real doctor or an intern, that it is at night when most people make love. It is night that dismantles the barricade to the heart.

Some people live out their whole lives within earshot of bells. As though it were the sound of those bells, the chiming, that were necessary to life, an essential element like oxygen. But you have chosen to live out your life within the sound of sirens, and page operators and cardiac monitors. To be on duty in an Emergency Room and to hear the sound of a siren is to listen to the gathering of a storm, or a battle growing nearby. Soon, you know, you will be fully engaged. It is a sound that, even now, after so many years, causes me to tremble. There will be days when you will regret that you chose sirens over bells. At such times you will hate the hospital with its dreadful Emergency Room and its Operating Room and its

Morgue. What a far cry from the snug lamplit hut of your boyhood dreams, with outside a raging snowstorm, and inside, a lovely fire in the grate. Here the storm is likely to whirl through the premises. The wolves know the way in, too. All the same, you will know that this hospital is your bones and your breath, without which you cannot live.

Tomorrow you will be an intern. Yours will be the glory of jotting down the conflicting orders of each one of your many superiors, and then carrying them out long after the rest have gone on to bed and other matters. You will be the tag end of Rounds, following, in single file, the First-Year Resident, the Second-Year Resident, the Third-Year Resident, the Chief Resident, and far ahead, caparisoned and plumed and glowing like Saint George—the Professor. And all the while, straining to hear their orders, gashing frantically at your notebook. Never mind. In four years, you, too, will become Chief Resident, second in line. To you, then, will fall the privilege of squiredom to your Professor. You shall carry his stethoscope, and you shall be to him dresser, clerk, guide, bandage, splint, scissors. You, his scrivener, his scullion, his squirm.

I must warn you that in every surgical training program the severest burden may not be the fatigue gained from working day and night, nor the fretfulness that comes from worrying over your patients, but the gauleiter mentality of some of the surgeons. These martinets infest every hospital to some degree, like rats in a silo. They are recognizable even from the time of their internships, when they can be observed preying upon medical students or barking impolitely at nurses. Invariably these surgeons rise to positions of great authority—Chief Resident, Professor of Surgery, Dean. Here, a malignant character can flourish unhampered. They are often superb technicians, each with unassailable credentials. To locate one of these surgeons, follow the sound of guffawing. You will come upon him, outside the room of a patient, surrounded by myrmidons, all taking their fun at the "crock" in the bed. Later, he will scold an elderly nurse for her unfamiliarity with the newest information revealed, if the truth be known, only the day before to him.

They are all big, good-looking fellows, given to wearing string ties and cowboy boots that can imply virility if need be. They own underdeveloped souls, the blighted wisps having slipped into their perfect frames at the moment of birth to live out their tenancy unacknowledged. These men tend to displace more air than they are allotted by body volume. Such a surgeon has a keen nose for misery, which he keeps plugged the way his forebears did their nostrils in a plague-struck city. Lest he be infected and suffer a pang of compassion. But observe him in the company of his superiors. Here he is all charm, all oiled congeniality at many decibels. Toward his peers he is wary; toward his intern, abusive or flippant.

Of these two, abuse or flippancy, you should hope for the former. Unlike flippancy, unvarnished cruelty does not interfere with the appetite, nor is it as likely to turn you murderous. Beware contagion from this surgeon. His pox are highly catching. Once so afflicted, you cannot be cured, but must depend ever after upon mere diligence and correctness to keep you out of the malpractice courts. Above all, do not yield to the temptation to punch him in the nose. Go instead to the bedside of one whose nightly fever is a hot red wind. Visit him again and see how the fever has receded beneath his skin to the deeper parts where it whines and spits and gathers itself for the next evening's eruption. See how in the morning he is rinsed of color like a blank page from which the most terrible words have been erased, but whose ghastly outline can still be seen. Do this, and you will have placed your cruel surgeon in his proper place. Still, there is redemption. Years later, I came upon an old Resident tormentor of mine. He was playing Mendelssohn on the violin. I watched his face take on the soft, unguarded tenderness of a man in love. And I marveled at the regenerative powers of the human spirit. But, then, Purgatory is on the way to Heaven, isn't it?

What! Your heart still dances toward Surgery? Why, then,

welcome to the Fellowship of the Knife. And God help you. If not
. . . if so much as one single seed of hesitation has fallen from my
pocket to germinate in your brain, turn aside. Let Surgery be a road
not taken and not rued. Only a flaming Torch reflected in water can
exist where it dare not go. Remember that Surgery, like Poetry, is a
subcelestial art. The angels disdain to perform either one of them.

Letter to a Young Surgeon

II

*At this, the start of your surgi-*cal internship, it is well that you be told how to behave in an operating room. You cannot observe decorum unless you first know what decorum is. Say that you have already changed into a scrub suit, donned cap, mask and shoe covers. You have scrubbed your hands and been helped into your gown and gloves. Now stand out of the way. Eventually, your presence will be noticed by the surgeon, who will motion you to take up a position at the table. Surgery is not one of the polite arts, as are Quilting and Illuminating Manuscripts. Decorum in the operating room does not include doffing your cap in the presence of nurses. Even the old-time surgeons knew this and operated without removing their hats.

The first rule of conversation in the operating room is silence. It is a rule to be broken freely by the Master, for he is engaged in the act of teaching. The forceful passage of bacteria through a face mask during speech increases the contamination of the wound and therefore the possibility of infection in that wound. It is a risk that must be taken. By the surgeon, wittingly, and by the patient, unbeknownst. Say what you will about a person's keeping control over his own destiny, there are some things that cannot be helped. Being made use of for teaching purposes in the operating room is one of them. It is an inevitable, admirable and noble circumstance. Besides, I have placated Fate too long to believe that She would bring on wound infection as the complication of such a high enterprise.

Observe the least movement of the surgeon's hands. See how he holds out his hand to receive the scalpel. See how the handle of it rides between his thumb and fingertips. The scalpel is the subtlest of the instruments, transmitting the nervous current in the surgeon's arm to the body of the patient. Too timidly applied, and it turns flabby, lifeless; too much pressure and it turns vicious. See how the surgeon applies the blade to the skin—holding it straight in its saddle lest he undercut and one edge of the incision be thinner than the other edge. The application of knife to flesh proclaims the master and exposes the novice. See the surgeon advancing his hand blindly into the abdomen as though it were a hollow in a tree. He is wary, yet needing to know. Will it be something soft and dead? Or a sudden pain in his bitten finger!

The point of the knife is called the *tang*, from the Latin word for *touch*. The sharp curving edge is the *belly* of the blade. The tang is for assassins, the belly for surgeons. Enough! You will not hold this knife for a long time. Do not be impatient for it. Nor reckon the time. Ripen only. Over the course of your training you will be given ever more elaborate tasks to perform. But for now, you must watch and wait. Excessive ego, arrogance and self-concern in an intern are out of place, as they preclude love for the patient on the table. There is no room for clever disobedience here. For the knife is like fire. The small child yearns to do what his father does, and he steals matches from the man's pocket. The fire he lights in his hiding place is beautiful to him; he toasts marshmallows in it. But he is just as likely to be burned. And reverence for the teacher is essential to the accumulation of knowledge. Even a bad surgeon will teach if only by the opportunity to see what not to do.

You will quickly come to detect the difference between a true surgeon and a mere product of the system. Democracy is not the best of all social philosophies in the selection of doctors for training in surgery. Anyone who so desires, and who is able to excel academically and who is willing to undergo the harsh training, can become a surgeon whether or not he is fit for the craft either manually or by temperament. If we continue to award licenses to the incompetent and the ill-suited, we shall be like those countries where work is given over not to those who can do it best, but to those who need it. That offers irritation enough in train stations; think of the result in airplane cockpits or operating rooms. Ponder long and hard upon this point. The mere decision to be a surgeon will not magically confer upon you the dexterity, compassion and calmness to do it.

Even on your first day in the operating room, you must look ahead to your last. An old surgeon who has lost his touch is like an old lion whose claws have become blunted, but not the desire to use them. Knowing when to quit and retire from the consuming passion of your life is instinctive. It takes courage to do it. But do it you must. No consideration of money, power, fame or fear of boredom may give you the slightest pause in laying down your scalpel when the first flagging of energy, bravery or confidence appears. To withdraw gracefully is to withdraw in a state of grace. To persist is to fumble your way to injury and ignominy.

Do not be dismayed by the letting of blood, for it is blood that animates this work, distinguishes it from its father, Anatomy. Red is the color in which the interior of the body is painted. If an operation be thought of as a painting in progress, and blood red the color on the brush, it must be suitably restrained and attract no undue attention; yet any insufficiency of it will increase the perishability of the canvas. Surgeons are of differing stripes. There are those who are slow and methodical, obsessive beyond all reason. These tortoises operate in a field as bloodless as a cadaver. Every speck of tissue in its proper place, every nerve traced out and brushed clean so that a Japanese artist could render it down to the dendrites. Should the contents of a single capillary be inadvertently shed, the whole procedure comes to a halt while Mr. Clean irrigates and suctions and mops and clamps and ties until once again the operative field looks like Holland at tulip time. Such a surgeon tells time not by the clock but by the calendar. For this, he is ideally equipped with an iron urinary bladder which he has disciplined to contract no more than once a day. To the drop-in observer, the work of such a surgeon is faultless. He gasps in admiration at the still life on the table. Should the same observer leave and return three hours later, nothing will have changed. Only a few more millimeters of perfection.

Then there are the swashbucklers who crash through the underbrush waving a machete, letting tube and ovary fall where they may. This surgeon is equipped with gills so that he can breathe under blood. You do not set foot in his room without a slicker and boots. Seasoned nurses quake at the sight of those arms, elbow-deep and *working*. It is said that one such surgeon entertained the other guests at a department Christmas party by splenectomizing a cat in thirty seconds from skin to skin.

Then there are the rest of us who are neither too timid nor too brash. We are just right. And now I shall tell you a secret. To be a good surgeon does not require immense technical facility. Compared to a violinist it is nothing. The Japanese artist, for one, is skillful at double brushing, by which technique he lays on color with one brush and shades it off with another, both brushes being held at the same time and in the same hand, albeit with different fingers. Come to think of it, a surgeon, like a Japanese artist, ought to begin his training at the age of three, learning to hold four or five instruments at a time in the hand while suturing with a needle and thread held in the teeth. By the age of five he would be able to dismantle and reconstruct an entire human body from calvarium to calcaneus unassisted and in the time it would take one of us to recite the Hippocratic Oath. A more obvious advantage of this baby surgeon would be his size. In times of difficulty he could be lowered whole into the abdomen. There, he could swim about, repair the works, then give three tugs on a rope and . . . Presto! Another gallbladder bites the dust.

In the absence of any such prodigies, each of you who is full-grown must learn to exist in two states—Littleness and Bigness. In your littleness you descend for hours each day through a cleft in the body into a tiny space that is both your workshop and your temple. Your attention in Lilliput is total and undistracted. Every artery is a river to be forded or dammed, each organ a mountain to be skirted or moved. At last, the work having been done, you ascend. You blink and look about at the vast space peopled by giants and massive furniture. Take a deep breath . . . and you are Big. Such instantaneous hypertrophy is the process by which a surgeon re-enters the outside world. Any breakdown in this resonance between the sizes causes the surgeon to live in a Renaissance painting where the depth perception is so bad.

Nor ought it to offend you that, a tumor having been successfully removed, and the danger to the patient having been circumvented, the very team of surgeons that only moments before had been a model of discipline and deportment comes loose at the seams and begins to wobble. Jokes are told, there is laughter, a hectic gaiety prevails. This is in no way to be taken as a sign of irreverence or callousness. When the men of the Kalahari return from the hunt with a haunch of zebra, the first thing everybody does is break out in a dance. It is a rite of thanksgiving. There will be food. They have made it safely home.

Man is the only animal capable of tying a square knot. During the course of an operation you may be asked by the surgeon to tie a knot. As drawing and coloring are the language of art, incising, suturing and knot tying are the grammar of surgery. A facility in knot tying is gained only by tying ten thousand of them. When the operation is completed, take home with you a package of leftover sutures. Light a fire in the fireplace and sit with your lover on a rug in front of the fire. Invite her to hold up her index finger, gently crooked in a gesture of beckoning. Using her finger as a strut, tie one of the threads about it in a square knot. Do this one hundred times. Now make a hundred grannies. Only then may you permit yourself to make love to her. This method of learning will not only enable you to master the art of knot tying, both grannies and square, it will bind you, however insecurely, to the one you love.

To do surgery without a sense of awe is to be a dandy—all style and no purpose. No part of the operation is too lowly, too menial. Even when suturing the skin at the end of a major abdominal procedure, you must operate with piety, as though you were embellishing a holy reliquary. The suturing of the skin usually falls to the lot of the beginning surgeon, the sights of the Assistant Residents and Residents having been firmly set upon more biliary, more gastric glories. In surgery, the love of inconsiderable things must govern your life—ingrown toenails, thrombosed hemorrhoids, warts. Never disdain the common ordinary ailment in favor of the exotic or rare. To the patient every one of his ailments is unique. One is not to be amused or captivated by disease. Only to a woodpecker is a wormy tree more fascinating than one uninhabited. There is only absorption in your patient's plight. To this purpose, willingly accept the smells and extrusions of the sick. To be spattered with the phlegm, vomitus and blood of suffering is to be badged with the highest office.

The sutured skin is all of his operation that the patient will see. It is your signature left upon his body for the rest of his life. For the patient, it is the emblem of his suffering, a reminder of his mortality. Years later, he will idly run his fingers along the length of the scar, and he will hush and remember. The good surgeon knows this. And so he does not overlap the edges of the skin, makes no dog-ears at the corners. He does not tie the sutures too tightly lest there be a row of permanent crosshatches. (It is not your purpose to construct a ladder upon which a touring louse could climb from pubis to navel and back.) The good surgeon does not pinch the skin with forceps. He leaves the proper distance between the sutures. He removes the sutures at the earliest possible date, and he uses sutures of the finest thread. All these things he does and does not do out of reverence for his craft and love for his patient. The surgeon who does otherwise ought to keep his hands in his pockets. At the end of the operation, cholecystectomy, say, the surgeon may ask you to slit open the gallbladder so that everyone in the room might examine the stones. Perform even this cutting with reverence as though the organ were still within the patient's body. You cut, and notice how the amber bile runs out, leaving a residue of stones. Faceted, shiny, they glisten. Almost at once, these wrested dewy stones surrender their warmth and moisture; they grow drab and dull. The descent from jewel to pebble takes place before your eyes.

Deep down, I keep the vanity that surgery is the red flower that blooms among the leaves and thorns that are the rest of Medicine. It is Surgery that, long after it has passed into obsolescence, will be remembered as the glory of Medicine. Then men shall gather in mead halls and sing of that ancient time when surgeons, like gods, walked among the human race. Go ahead. Revel in your Specialty; it is your divinity.

It is quest and dream as well:

The incision has been made. One expects mauve doves and colored moths to cloud out of the belly in celebration of the longed-for coming. Soon the surgeon is greeted by the eager blood kneeling and offering its services. Tongues of it lap at his feet; flames and plumes hold themselves aloft to light his way. And he follows this guide that flows just ahead of him through rifts, along the edges of cliffs, picking and winding, leaping across chasms, at last finding itself and pooling to wait for him. But the blood cannot wait a moment too long lest it become a blob of coagulum, something annulled by its own puddling. The surgeon rides the patient, as though he were riding a burro down into a canyon. This body is beautiful to him, and he to it—he whom the patient encloses in the fist of his flesh. For months, ever since the first wild mitosis, the organs had huddled like shipwrecks. When would he come? Will he never come? And suddenly, into the sick cellar—fingers of light! The body lies stupefied at the moment of encounter. The cool air stirs the buried flesh. Even the torpid intestine shifts its slow coils to make way.

Now the surgeon must take care. The fatal glissade, once begun, is not to be stopped. Does this world, too, he wonders, roll within the precincts of mercy? The questing dreamer leans into the patient to catch the subtlest sounds. He hears the harmonies of their two bloods, his and the patient's. They sing of death and the beauty of the rose. He hears the playing together of their two breaths. If Pythagoras is right, there is no silence in the universe. Even the stars make music as they move.

Only do not succumb to self-love. I know a surgeon who, having left the room, is certain, beyond peradventure of doubt, that his disembodied radiance lingers on. And there are surgeons of such aristocratic posture that one refrains only with difficulty from slipping them into the nobility. As though they had risen from Mister to Doctor to Professor, then on to Baron, Count, Archduke, then further, to Apostle, Saint. I could go further.

Such arrogance can carry over to the work itself. There was a surgeon in New Haven, Dr. Truffle, who had a penchant for long midline incisions—from sternum to pubis—no matter the need for exposure. Somewhere along the way, this surgeon had become annoyed by the presence of the navel, which, he decided, interrupted the pure line of his slice. Day in, day out, it must be gone around, either to the right or to the left. Soon, what was at first an annoyance became a hated impediment that must be got rid of. Mere circumvention was not enough. And so, one day, having arrived at the midpoint of his downstroke, this surgeon paused to cut out the navel with a neat ellipse of skin before continuing on down to the pubis. Such an elliptical incision when sutured at the close of the operation forms the continuous straight line without which this surgeon could not live. Once having cut out a navel (the first incidental umbilectomy, I suppose, was the hardest) and seeing the simple undeviate line of his closure, he vowed never again to leave a navel behind. Since he was otherwise a good surgeon, and very successful, it was not long before there were thousands of New Haveners walking around minus their belly buttons. Not that this interfered with any but the most uncommon of activities, but to those of us who examined them postoperatively, these abdomens had a blind, bland look. Years later I would happen upon one of these bellies and know at once the author of the incision upon it. Ah, I would say, Dr. Truffle has been here.

It is so difficult for a surgeon to remain "unconscious," retaining the clarity of vision of childhood, to know and be secure in his ability, yet be unaware of his talents. It is almost impossible. There are all too many people around him paying obeisance, pandering, catering, beaming, lusting. Yet he must try.

It is not enough to love your work. Love of work is a kind of self-indulgence. You must go beyond that. Better to perform endlessly, repetitiously, faithfully, the simplest acts, like trimming the toenails of an old man. By so doing, you will not say *Here I Am*, but *Here It Is*. You will not announce your love but will store it up in the bodies of your patients to carry with them wherever they go.

Many times over, you will hear otherwise sensible people say, "You have golden hands," or, "Thanks to you and God, I have recovered." (Notice the order in which the credit is given.) Such ill-directed praise has no significance. It is the patient's disguised expression of relief at having come through, avoided death. It is a private utterance, having nothing to do with you. Still, such words are enough to turn a surgeon's head, if any more turning were needed.

Avoid these blandishments at all cost. You are in service to your patients, and a servant should know his place. The world is topsy-turvy in which a master worships his servant. You are a kindly, firm, experienced servant, but a servant still. If any patient of mine were to attempt to bathe my feet, I'd kick over his basin, suspecting that he possessed not so much a genuine sentiment as a conventional one. It is beneath your dignity to serve as an object of veneration or as the foil in an act of contrition. To any such effusion a simple "Thank you" will do. The rest is pride, and everyone knoweth before *what* that goeth.

Alexander the Great had a slave whose sole responsibility was to whisper "Remember, you are mortal" when he grew too arrogant. Perhaps every surgeon should be assigned such a deflator. The surgeon is the mere instrument which the patient takes in his hand to heal himself. An operation, then, is a time of revelation, both physical and spiritual, when, for a little while, the secrets of the body are set forth to be seen, to be touched, and the surgeon himself is laid open to Grace.

An operation is a reenactment of the story of Jonah and the Whale. In surgery, the patient is the whale who swallows up the surgeon. Unlike Jonah, however, the surgeon does not cry out *non serviam,* but willingly descends into the sick body in order to cut out of it the part that threatens to kill it. In an operation where the patient is restored to health, the surgeon is spewed out of the whale's body, and both he and his patient are healed. In an operation where the patient dies on the table, the surgeon, although he is rescued from the whale and the sea of blood, is not fully healed, but will bear the scars of his sojourn in the belly of the patient for the rest of his life.

Letter to a Young Surgeon
III

All right. You fainted in the operating room, had to go sit on the floor in a corner and put your head down. You are making altogether too much of it. You have merely announced your humanity. Only the gods do not faint at the sight of the *mysterium tremendum*; they have too jaded a glance. At the same place in the novitiate I myself more than once slid ungracefully to the floor in the middle of things. It is less a sign of weakness than an expression of guilt. A flinching in the face of the forbidden.

The surgeon is an explorer in the tropical forest of the body. Now and then he reaches up to bring closer one of the wondrous fruits he sees there. Before he departs this place, he knows that he must pluck one of them. He knows, too, that it is forbidden to do so. But it is a trophy, no, a *spoil* that has been demanded of him by his patron, the patient, who has commissioned and outfitted him for this exploration. At last the surgeon holds the plucked organ in his hand, but he is never wholly at ease. For what man does not grow shy, fearful, before the occult uncovered?

Don't worry. The first red knife is the shakiest. This is as true for the assassin as it is for the surgeon. The assassin's task is easier, for he is more likely to be a fanatic. And nothing steadies the hand like zeal. The surgeon's work is madness icily reined in to a good purpose. Still I know that it is perverse to relieve pain by inflicting it. This requires that the patient give over to you his free will and his trust. It is too much to ask. Yet we do every day, and with the arrogance born of habit and custom, and grown casual, even charming.

"Come, lie down on this table," you say, and smile. Your voice is soft and reasonable.

"Where will you cut me open?" the patient asks. And he grips his belly as though he and it were orphan twins awaiting separate adoption. The patient's voice is *not* calm; it trembles and quavers.

"From here . . . to here," you reply, and you draw a fingernail across his shuddering flesh. His navel leaps like a flushed bird. Oh, God! He has heard the knell of disembowelment. You really *do* mean to do it! And you do, though not with the delectation that will be attributed to you by those who do not do this work.

The cadaver toward which I have again and again urged you is like an abundant nest from which the birds have long since flown. It is a dry, uninhabited place—already dusty. It is a "thing" that the medical student will pull apart and examine, seeking evidence, clues from which he can reconstruct the life that once flourished there. The living patient is a nest in which a setting bird huddles. She quivers, but does not move when you press aside leaves in order to see better. Your slightest touch frightens her. You hold your breath and let the leaves spring back to conceal her. You want so much for her to trust you.

In order to do good works throughout his lifetime, a man must strive ever higher to carry out his benefices; he must pray, defer pleasure and steel himself against temptation. And against fainting. The committing of surgery grows easier and easier, it seems, until the practice is second nature. Come, come! You fainted! Why don't you admit that you are imperfect, and that you strain to appeal to yourself and to others? Surgery is, in one sense, a judicious contrivance, like poetry. But . . . it is an elect life, here among the ranting machinery and brazen lamps of the operating room, where on certain days now the liquidity of the patient reminds me of the drought that is attacking my own flesh. Listen, I will tell you what you already know: There is nothing like an honest piece of surgery. Say what you will, there is nothing more satisfying to the spirit than . . . the lancing of a boil!

Behold the fierce, hot protuberance compressing the howling nerves about it. You sound it with your fingers. A light tap brings back a malevolent answering wave, and a groan from the patient. Now the questing knife rides your hand. Go! And across the swelling gallops a thin red line. Again! Deeper, plowing. All at once there lifts a wave of green mud. Suddenly the patient's breathing comes more easily, his tense body relaxes, he smiles. For him it is like being touched by the hand of God. It is a simple act, requiring not a flicker of intellect nor a whisker of logic. To outwit disease it takes a peasant's cunning, not abstract brilliance. It is like the felling of a tree for firewood. Yet not poetry, nor music nor mathematics can bring such gladness, for riding out upon that wave of pus has come the black barque of pain. Just so will you come to love the boils and tumors of your patients.

I shall offer you two antidotes to fainting in the operating room.

1. Return as often as possible to the Anatomy Laboratory. As the sculptor must gain unlimited control over his marble, the surgeon must "own" the flesh. As drawing is to the painter, so is anatomy to the surgeon. You must continue to dissect for the rest of your life. To raise a flap of skin, to trace out a nerve to its place of confluence, to carry a tendon to its bony insertion, these are things of grace and beauty. They are simple, nontheoretical, workaday acts which, if done again and again, will give rise to that profound sense of structure that is the birthplace of intuition.

It is only at the dissecting table that you can find the models of your art. Only there that you will internalize the structure and form of the body so that any variations or anomalies or unforeseen circumstances are not later met with dismay and surprise. Unlike the face, the internal organs bear a remarkable sameness to one another. True, there are differences in the size of normal kidneys, livers and spleens, and there are occasional odd lobulations and unusual arrangements of ducts and vessels, but by and large, one liver is very like another. A kidney is a kidney. Unlike a face, it bears no distinctive mark or expression that would stamp it as the kidney of Napoleon Bonaparte, say, or Herman Melville. It is this very sameness that makes of surgery a craft that can be perfected by repetition and industry. Therefore, return to the Anatomy Laboratory. Revere and follow your prosector. The worship and awe you show the cadavers will come back to you a thousandfold. Even now, such an old knife as I goes to that place to dissect, to probe, to delve. What is an operating room but a prosectorium that has been touched into life?

2. Do not be impatient to wield the scalpel. To become a surgeon is a gradual, imperceptible, subtle transformation. Do not hurry from the side of the one who instructs you, but stay with your "master" until he bids you to go. It is his office to warm you with his words, on rounds and in the operating room, to color the darkness and shade the brilliance of light until you have grown strong enough to survive. Then, yes, leave him, for no sapling can grow to fullness in the shade of a big tree.

Do these things that I have told you and you will not faint in the operating room. I do not any longer faint, nor have I for thirty years. But now and then, upon leaving the hospital after a long and dangerous operation has been brought to a successful close, I stroke the walls of the building as though it were a faithful animal that has behaved itself well.

Brute

You must never again set your anger upon a patient. You were tired, you said, and therefore it happened. Now that you have excused yourself, there is no need for me to do it for you.

Imagine that you yourself go to a doctor because you have chest pain. You are worried that there is something the matter with your heart. Chest pain is your Chief Complaint. It happens that your doctor has been awake all night with a patient who has been bleeding from a peptic ulcer of his stomach. He is tired. That is your doctor's Chief Complaint. I have chest pain, you tell him. I am tired, he says.

Still I confess to some sympathy for you. I know what tired is.

Listen: It is twenty-five years ago in the Emergency Room. It is two o'clock in the morning. There has been a day and night of stabbings, heart attacks and automobile accidents. A commotion at the door: A huge black man is escorted by four policemen into the Emergency Room. He is handcuffed. At the door, the man rears as though to shake off the men who cling to his arms and press him from the rear. Across the full length of his forehead is a laceration. It is deep to the bone. I know it even without probing its depths. The split in his black flesh is like the white wound of an ax in the trunk of a tree. Again and again he throws his head and shoulders forward, then back, rearing, roaring. The policemen ride him like parasites. Had he horns he would gore them. Blind and trussed, the man shakes them about, rattles them. But if one of them loses his grip, the others are still fixed and sucking. The man is hugely drunk —toxic, fuming, murderous—a great mythic beast broken loose in the city, surprised in his night raid by a phalanx of legionnaires armed with clubs and revolvers.

I do not know the blow that struck him on the brow. Or was there any blow? Here is a brow that might have burst on its own, spilling out its excess of rage, bleeding itself toward ease. Perhaps it was done by a jealous lover, a woman, or a man who will not pay him the ten dollars he won on a bet, or still another who has hurled the one insult that he cannot bear to hear. Perhaps it was done by the police themselves. From the distance of many years and from the safety of my little study, I choose to see it thus:

The helmeted corps rounds the street corner. A shout. "There he is!" And they clatter toward him. He stands there for a moment, lurching. Something upon which he had been feeding falls from his open mouth. He turns to face the policemen. For him it is not a new challenge. He is scarred as a Zulu from his many battles. Almost from habit he ascends to the combat. One or more of them falls under his flailing arms until—there is the swing of a truncheon, a sound as though a melon has been dropped from a great height. The white wedge appears upon the sweating brow of the black man, a waving fall of blood pours across his eyes and cheeks.

The man is blinded by it; he is stunned. Still he reaches forth to make contact with the enemy, to do one more piece of damage. More blows to the back, the chest and again to the face. Bloody spume flies from his head as though lifted by a great wind. The police are spattered with it. They stare at each other with an abstract horror and disgust. One last blow, and, blind as Samson, the black man undulates, rolling in a splayfooted circle. But he does not go down. The police are upon him then, pinning him, cuffing his wrists, kneeing him toward the van. Through the back window of the wagon—a netted panther.

In the Emergency Room he is led to the treatment area and to me. There is a vast dignity about him. He keeps his own counsel. What is he thinking? I wonder. The police urge him up on the table. They put him down. They restrain his arms with straps. I examine the wound, and my heart sinks. It is twelve centimeters long, irregular, jagged and, as I knew, to the skull. It will take at least two hours.

I am tired. Also to the bone. But something else . . . Oh, let me not deny it. I am ravished by the sight of him, the raw, untreated flesh, his very wildness which suggests less a human than a great and beautiful animal. As though by the addition of the wound, his body is more than it was, more of a body. I begin to cleanse and debride the wound. At my touch, he stirs and groans. "Lie still," I tell him. But now he rolls his head from side to side so that I cannot work. Again and again he lifts his pelvis from the table, strains against his bonds, then falls heavily. He roars something, not quite language. "Hold still," I say. "I cannot stitch your forehead unless you hold still."

Perhaps it is the petulance in my voice that makes him resume his struggle against all odds to be free. Perhaps he understands that it is only a cold, thin official voice such as mine, and not the billy clubs of half-a-dozen cops that can rob him of his dignity. And so he strains and screams. But why can he not sense that I am tired? He spits and curses and rolls his head to escape from my fingers. It is quarter to three in the morning. I have not yet begun to stitch. I lean close to him; his steam fills my nostrils. "Hold still," I say.

"*You* fuckin' hold still," he says to me in a clear, fierce voice. Suddenly, I am in the fury with him. Somehow he has managed to capture me, to pull me inside his cage. Now we are two brutes hissing and batting at each other. But I do not fight fairly.

I go to the cupboard and get from it two packets of heavy, braided silk suture and a large curved needle. I pass one of the heavy silk sutures through the eye of the needle. I take the needle in the jaws of a needle holder, and I pass the needle through the center of his right earlobe. Then I pass the needle through the mattress of the stretcher. And I tie the thread tightly so that his head is pulled to the right. I do exactly the same to his left earlobe, and again I tie the thread tightly so that his head is facing directly upward.

"I have sewn your ears to the stretcher," I say. "Move, and you'll rip 'em off." And leaning close I say in a whisper, "Now *you* fuckin' hold still."

I do more. I wipe the gelatinous clots from his eyes so that he can see. And I lean over him from the head of the table, so that my face is directly above his, upside down. And I grin. It is the cruelest grin of my life. Torturers must grin like that, beheaders and operators of racks.

But now he does hold still. Surely it is not just fear of tearing his earlobes. He is too deep into his passion for that. It is more likely some beastly wisdom that tells him that at last he has no hope of winning. That it is time to cut his losses, to slink off into high grass. Or is it some sober thought that pierces his wild brain, lacerating him in such a way that a hundred nightsticks could not? The thought of a woman who is waiting for him, perhaps? Or a child who, the next day and the week after that, will stare up at his terrible scars with a silent wonder that will shame him? For whatever reason, he is perfectly still.

It is four o'clock in the morning as I take the first stitch in his wound. At five-thirty, I snip each of the silks in his earlobes. He is released from his leg restrainers and pulled to a sitting position. The bandage on his head is a white turban. A single drop of blood in each earlobe, like a ruby. He is a maharajah.

The police return. All this time they have been drinking coffee with the nurses, the orderlies, other policemen, whomever. For over three hours the man and I have been alone in our devotion to the wound. "I have finished," I tell them. Roughly, they haul him from the stretcher and prod him toward the door. "Easy, easy," I call after them. And, to myself, if you hit him again . . .

Even now, so many years later, this ancient rage of mine returns to peck among my dreams. I have only to close my eyes to see him again wielding his head and jaws, to hear once more those words at which the whole of his trussed body came hurtling toward me. How sorry I will always be. Not being able to make it up to him for that grin.

Toenails

———◆•◆———

It is the custom of many doctors,
I among them, to withdraw from the practice of medicine every
Wednesday afternoon. This, only if there is no patient who de-
mands the continuous presence of his physician. I urge you, when
the time comes, to do it, too. Such an absence from duty ought not
to win you the accusation of lèse responsibility. You will, of course,
have secured the availability of a colleague to look after your patients for the few hours you will spend grooming and watering
your spirit. Nor is such idleness a reproach to those who do not
take time off from their labors, but who choose to scramble on
without losing the pace. Loafing is not better than frenzied determi-
nation. It is but an alternate mode of living.

Long ago I made a vow that I would never again delve away
the month of July in the depths of the human body. In July it would
be my own cadaver that engaged me. There is a danger in becoming
too absorbed in Anatomy. At the end of eleven months of dissec-

tion, you stand in fair risk of suffering a kind of rapture of the deep, wherein you drift, tumbling among the coils of intestine in a state of helpless enchantment. Only a month's vacation can save you. It is wrongheaded to think of total submersion in the study and practice of Medicine. That is going too far. And going too far is for saints. I know medical students well enough to exclude you from that slender community.

Nor must you be a priest who does nothing but preserve the souls of his parishioners and lets his own soul lapse. Such is the burnt-out case who early on drinks his patients down in a single radiant gulp and all too soon loses the desire to practice Medicine at all. In a year or two he is to be found lying in bed being fed oatmeal with a spoon. Like the fruit of the Amazon he is too quickly ripe and too quickly rotten.

Some doctors spend Wednesday afternoon on the golf course. Others go fishing. Still another takes a lesson on the viola da gamba. I go to the library where I join that subculture of elderly men and women who gather in the Main Reading Room to read or sleep beneath the world's newspapers, and thumb through magazines and periodicals, educating themselves in any number of esoteric ways, or just keeping up. It is not the least function of a library to provide for these people a warm, dry building with good working toilets and, ideally, a vending machine from which to buy a cup of hot broth or coffee. All of which attributes a public library shares with a neighborhood saloon, the only difference being the beer of one and the books of the other.

How brave, how reliable they are! plowing through you-name-what inclemencies to get to the library shortly after it opens. So unbroken is their attendance that, were one of them to be missing, it would arouse the direst suspicions of the others. And of me. For I have, furtively at first, then with increasing recklessness, begun to love them. They were, after all, living out my own fantasies. One day, with luck, I, too, would become a full-fledged, that is to say *daily*, member. At any given time, the tribe consists of a core of six regulars and a somewhat less constant pool of eight others of whom two or three can be counted on to appear. On very cold days, all eight of these might show up, causing a bit of a jam at the newspaper rack, and an edginess among the regulars.

Either out of loyalty to certain beloved articles of clothing, or from scantiness of wardrobe, they wear the same things every day. For the first year or two this was how I identified them. Old Stovepipe, Mrs. Fringes, Neckerchief, Galoshes—that sort of thing. In no other society does apparel so exactly fit the wearer as to form a part of his persona. Dior, Balenciaga, take heed! By the time I arrive, they have long since devoured the morning's newspapers and settled into their customary places. One or two, Galoshes, very likely, and Stovepipe, are sleeping it off. These two seem to need all the rest they can get. Mrs. Fringes, on the other hand, her hunger for information unappeasable, having finished all of the newspapers, will be well into the *Journal of Abnormal Psychology*, the case histories of which keep her riveted until closing time. As time went by, despite that we had not yet exchanged a word of conversation, I came to think of them as dear colleagues, fellow readers who, with me, were engaged in the pursuit of language. Nor, I noticed, did they waste much time speaking to each other. Reading was serious business. Only downstairs near the basement vending machine would animated conversation break out. Upstairs, in the Reading Room, the vow of silence was sacred.

I do not know by what criteria such selections are made, but Neckerchief is my favorite. He is a man well into his eighties with the kind of pink face that even in July looks as though it has just been brought in out of the cold. A single drop of watery discharge, like a crystal bead, hangs at the tip of his nose. His gait is stiff-legged, with tiny, quick, shuffling steps accompanied by rather wild arm-swinging in what seems an effort to gain momentum or maintain balance. For a long time I could not decide whether this manner of walking was due to arthritis of the knees or to the fact that for most of the year he wore two or more pairs of pants. Either might have been the cause of his lack of joint flexion. One day, as I held the door to the Men's Room for him, he pointed to his knees and announced, by way of explanation of his slowness:

"The hinges is rusty."

The fact was delivered with a shake of the head, a wry smile and without the least self-pity.

"No hurry," I said, and once again paid homage to Sir William Osler, who instructed his students to "listen to the patient. He is trying to tell you what is the matter with him." From that day, Neckerchief and I were friends. I learned that he lives alone in a rooming house eight blocks away, that he lives on his Social Security check, that his wife died a long time ago, that he has no children, and that the *Boston Globe* is the best damn newspaper in the library. He learned approximately the same number of facts about me. Beyond that we talked about politics and boxing, which is his great love. He himself had been an amateur fighter sixty years ago—most of his engagements having been spontaneous brawls of a decidedly ethnic nature. "It was the Polacks against the Yids," he told me, "and both of 'em against the Micks." He held up his fists to show the ancient fractures.

The actual neckerchief is a classic red cowboy rag folded into a triangle and tied about his neck in such a way that the widest part lies at the front, covering the upper chest as a kind of bib. Now and then a nose drop elongates, shimmers, wobbles and falls to be absorbed into the neckerchief. Meanwhile a new drop has taken the place of the old. So quickly is this newcomer born that I, for one, have never beheld him unadorned.

One day I watched as Neckerchief, having raided the magazine rack, journeyed back to his seat. In one flapping hand the *Saturday Review* rattled. As he passed, I saw that his usually placid expression was replaced by the look of someone in pain. Each step was a fresh onslaught of it. His lower lip was caught between his teeth. His forehead had been cut and stitched into lines of endurance. He was hissing. I waited for him to take his seat, which he did with a gasp of relief, then went up to him.

"The hinges?" I whispered.

"Nope. The toes."

"What's wrong with your toes?"

"The toenails is too long. I can't get at 'em. I'm walkin' on 'em."

I left the library and went to my office.

"What are you doing here?" said my nurse. "It's Wednesday afternoon. People are just supposed to die on Wednesday afternoon."

"I need the toenail cutters. I'll bring them back tomorrow."

"The last time you took something out of here I didn't see it for six months."

Neckerchief was right where I had left him. A brief survey, however, told me that he had made one trek in my absence. It was *U.S. News & World Report* on his lap. The *Saturday Review* was back in the rack. I could only guess what the exchange had cost him. I doubted that either of the magazines was worth it.

"Come on down to the Men's Room," I said. "I want to cut your toenails." I showed him my toenail clippers, the heavy-duty kind that you grip with the palm, and with jaws that could bite through bone. One of the handles is a rasp. I gave him a ten-minute head start, then followed him downstairs to the Men's Room. There was no one else there.

"Sit here." I pointed to one of the booths. He sat on the toilet. I knelt and began to take off his shoes.

"Don't untie 'em," he said. "I just slide 'em on and off."

The two pairs of socks were another story, having to be peeled off. The underpair snagged on the toenails. Neckerchief winced.

"How do you get these things on?" I asked.

"A mess, ain't they? I hope I don't stink too bad for you."

The nail of each big toe was the horn of a goat. Thick as a thumb and curved, it projected down over the tip of the toe to the underside. With each step, the nail would scrape painfully against the ground and be pressed into his flesh. There was dried blood on each big toe.

"Jesus, man!" I said. "How can you walk?" I thought of the eight blocks he covered twice a day.

It took an hour to do each big toe. The nails were too thick even for my nail cutters. They had to be chewed away little by little, then flattened out with the rasp. Now and then a fragment of nail would fly up, striking me in the face. The other eight toes were easy. Now and then, the door opened. Someone came and went to the row of urinals. Twice, someone occupied the booth next to ours. I never once looked up to see. They'll just have to wonder, I thought. But Neckerchief could tell from my face.

"It doesn't look decent," he said.

"Never mind," I told him. "I bet this isn't the strangest thing that's happened down here." I wet some toilet paper with warm water and soap, washed each toe, dried him off, and put his shoes and socks back on. He stood up and took a few steps, like someone who is testing the fit of a new pair of shoes.

"How is it?"

"It don't hurt," he said, and gave me a smile that I shall keep in my safety-deposit box at the bank until the day I die.

"That's a Cadillac of a toe job," said Neckerchief. "How much do I owe ya?"

"On the house," I said. "And besides, what kind of a boy do you think I am?"

The next week I did Stovepipe. He was an easy case. Then, Mrs. Fringes, who was a special problem. I had to do her in the Ladies' Room, which tied up the place for half an hour. A lot of people opened the door, took one look, and left in a hurry. Either it was hot in there or I had a temperature.

I never go to the library on Wednesday afternoon without my nail clippers in my briefcase. You just never know.

Mercy

It is October at the Villa Serbelloni, where I have come for a month to write. On the window ledges the cluster flies are dying. The climate is full of uncertainty. Should it cool down? Or warm up? Each day it overshoots the mark, veering from frost to steam. The flies have no uncertainty. They understand that their time has come.

What a lot of energy it takes to die! The frenzy of it. Long after they have collapsed and stayed motionless, the flies are capable of suddenly spinning so rapidly that they cannot be seen. Or seen only as a blurred glitter. They are like dervishes who whirl, then stop, and lay as quiet as before, only now and then waving a leg or two feebly, in a stuporous reenactment of locomotion. Until the very moment of death, the awful buzzing as though to swarm again.

Every morning I scoop up three dozen or so corpses with a dustpan and brush. Into the wastebasket they go, and I sit to begin the day's writing. All at once, from the wastebasket, the frantic knocking of resurrection. Here, death has not yet secured the premises. No matter the numbers slaughtered, no matter that the windows be kept shut all day, each evening the flies gather on the ledges to die, as they have lived, *ensemble*. It must be companionable to die so, matching spin for spin, knock for knock, and buzz for buzz with one's fellows. We humans have no such fraternity, but each of us must buzz and spin and knock alone.

I think of a man in New Haven! He has been my patient for seven years, ever since the day I explored his abdomen in the operating room and found the surprise lurking there—a cancer of the pancreas. He was forty-two years old then. For this man, these have been seven years of famine. For his wife and his mother as well. Until three days ago his suffering was marked by slowly increasing pain, vomiting and fatigue. Still, it was endurable. With morphine. Three days ago the pain rollicked out of control, and he entered that elect band whose suffering cannot be relieved by any means short of death. In his bed at home he seemed an eighty-pound concentrate of pain from which all other pain must be made by serial dilution. He twisted under the lash of it. An ambulance arrived. At the hospital nothing was to be done to prolong his life. Only the administration of large doses of narcotics.

"Please," he begs me. In his open mouth, upon his teeth, a brown paste of saliva. All night long he has thrashed, as though to hollow out a grave in the bed.

"I won't let you suffer," I tell him. In his struggle the sheet is thrust aside. I see the old abandoned incision, the belly stuffed with tumor. His penis, even, is skinny. One foot with five blue toes is exposed. In my cupped hand, they are cold. I think of the twenty bones of that foot laced together with tendon, each ray accompanied by its own nerve and artery. Now, this foot seems a beautiful dead animal that had once been trained to transmit the command of a man's brain to the earth.

"I'll get rid of the pain," I tell his wife.

But there is no way to kill the pain without killing the man who owns it. Morphine to the lethal dose . . . and still he miaows and bays and makes other sounds like a boat breaking up in a heavy sea. I think his pain will live on long after he dies.

"Please," begs his wife, "we cannot go on like this."

"Do it," says the old woman, his mother. "Do it now."

"To give him any more would kill him," I tell her.

"Then do it," she says. The face of the old woman is hoof-beaten with intersecting curves of loose skin. Her hair is donkey brown, donkey gray.

They wait with him while I go to the nurses' station to prepare the syringes. It is a thing that I cannot ask anyone to do for me. When I return to the room, there are three loaded syringes in my hand, a rubber tourniquet and an alcohol sponge. Alcohol sponge! To prevent infection? The old woman is standing on a small stool and leaning over the side rail of the bed. Her bosom is just above his upturned face, as though she were weaning him with sorrow and gentleness from her still-full breasts. All at once she says severely, the way she must have said it to him years ago:

"Go home, son. Go home now."

I wait just inside the doorway. The only sound is a flapping, a rustling, as in a room to which a small animal, a bat perhaps, has retreated to die. The women turn to leave. There is neither gratitude nor reproach in their gaze. I should be hooded.

At last we are alone. I stand at the bedside.

"Listen," I say, "I can get rid of the pain." The man's eyes regain their focus. His gaze is like a wound that radiates its pain outward so that all upon whom it fell would know the need of relief.

"With these." I hold up the syringes.

"Yes," he gasps. "Yes." And while the rest of his body stirs in answer to the pain, he holds his left, his acquiescent arm still for the tourniquet. An even dew of sweat covers his body. I wipe the skin with the alcohol sponge, and tap the arm smartly to bring out the veins. There is one that is still patent; the others have long since clotted and broken down. I go to insert the needle, but the tourniquet has come unknotted; the vein has collapsed. Damn! Again I tie the tourniquet. Slowly the vein fills with blood. This time it stays distended.

He reacts not at all to the puncture. In a wild sea what is one tiny wave? I press the barrel and deposit the load, detach the syringe from the needle and replace it with the second syringe. I send this home, and go on to the third. When they are all given, I pull out the needle. A drop of blood blooms on his forearm. I blot it with the alcohol sponge. It is done. In less than a minute, it is done.

"Go home," I say, repeating the words of the old woman. I turn off the light. In the darkness the contents of the bed are theoretical. No! I must watch. I turn the light back on. How reduced he is, a folded parcel, something chipped away until only its shape and a little breath are left. His impatient bones gleam as though to burst through the papery skin. I am impatient, too. I want to get it over with, then step out into the corridor where the women are waiting. His death is like a jewel to them.

My fingers at his pulse. The same rhythm as mine! As though there were one pulse that beat throughout all of nature, and every creature's heart throbbed precisely.

"You can go home now," I say. The familiar emaciated body untenses. The respirations slow down. Eight per minute . . . six . . . It won't be long. The pulse wavers in and out of touch. It won't be long.

"Is that better?" I ask him. His gaze is distant, opaque, preoccupied. Minutes go by. Outside, in the corridor, the murmuring of women's voices.

But this man will not die! The skeleton rouses from its stupor. The snout twitches as if to fend off a fly. What is it that shakes him like a gourd full of beans? The pulse returns, melts away, comes back again, and stays. The respirations are twelve, then fourteen. I have not done it. I did not murder him. I am innocent!

I shall walk out of the room into the corridor. They will look at me, holding their breath, expectant. I lift the sheet to cover him. All at once, there is a sharp sting in my thumb. The same needle with which I meant to kill him has pricked *me*. A drop of blood appears. I press it with the alcohol sponge. My fresh blood deepens the stain of his on the gauze. Never mind. The man in the bed swallows. His Adam's apple bobs slowly. It would be so easy to do it. Three minutes of pressure on the larynx. He is still not conscious, wouldn't feel it, wouldn't know. My thumb and fingertips hover, land on his windpipe. My pulse beating in his neck, his in mine. I look back over my shoulder. No one. Two bare IV poles in a corner, their looped metal eyes witnessing. Do it! Fingers press. Again he swallows. Look back again. How closed the door is. And . . . my hand wilts. I cannot. It is not in me to do it. Not that way. The man's head swivels like an upturned fish. The squadron of ribs battles on.

I back away from the bed, turn and flee toward the doorway. In the mirror, a glimpse of my face. It is the face of someone who has been resuscitated after a long period of cardiac arrest. There is no spot of color in the cheeks, as though this person were in shock at what he had just seen on the yonder side of the grave.

In the corridor the women lean against the wall, against each other. They are like a band of angels dispatched here to take possession of his body. It is the only thing that will satisfy them.

"He didn't die," I say. "He won't . . . or can't." They are silent.

"He isn't ready yet," I say.

"He *is* ready," the old woman says. "*You* ain't."

Rounds

I hereby call for a moratorium on the hideous custom of naming a hospital *Memorial* something. Or something *Memorial*. As in Veterans' Memorial Hospital or Massachusetts Memorial Hospital or just plain Memorial Hospital. The word has too mortuary an odor. In such a crepuscular place the doctor himself is a shade who has only his familiarity with pain to offer. Here, there can be no hope—only the fulfillment of death. O give me happier hospitals: Veterans' Recovery Hospital or the Massachusetts Palace of Healing, or the Strong Comeback Clinic where, should a patient fail to get well, it would be an injustice rather than a foregone conclusion.

Sometimes, in the Memorial Hospital where I work, I feel like a survivor wandering dazed through burning streets. I pause to study a corpse splayed upon the tines of a great gate. Was he, I wonder, looter or defender? And then I hear the ghostly moan of one not yet dead of his wounds, but still calling out in the language of the living. At the end of just such a night I went to the Maternity floor (I had no business there) and stood in the corridor outside the Delivery Room. Through the closed doors I listened and heard the first cry of a fellow human being. When I heard the newcomer . . . miracle! I was healed.

I understand your reluctance to leave your ward and move on to the next Service. I, too, suffered such dislocations. There is a certain melancholy one feels in passing through a ward full of new patients about whom one knows nothing but for whom one is to be responsible. It is the same sadness a forest ranger must feel in the contemplation of bonsai whose fate it is to be bound, twisted and amputated into facsimiles of healthy trees. But dwell among these patients for a day or two, become familiar with their wounds and their pain, and all at once you feel a surge of happiness. It is the gift of the patients to you—your knighthood conferred upon you by them. Now you are in a position for healing.

Within days, this new ward is like a village of friends into which you have stumbled and to which you are attached by chains of trust. These patients would risk everything, their lives even, to save you. You can see this from the little smiles they let play upon you when you pass. And you will love them to the point of pain and beyond. But loving them is not enough. To be of real help you must wish to prolong by one second some comfortable moment of their lives. To do this, a doctor must be able to do what even he cannot imagine doing.

Let us go then, you and I, to make Rounds together in this Memorial Hospital. It is a companionable thing for doctors to do.

ROOM ONE

A man carries a small bouquet of violets into a room where a woman lies sleeping. His feet both hurry and drag at the same time. As though his right foot were in desperate haste, and his left were reluctant. This man is not used to holding flowers. He is a plumber; his hands are stained with grease, and scarred. The violets seem even more fragile in his great burry fist. Now the man leans toward the bed, watching the face of the woman. His own face wilts under the onslaught of her labored breath. He stuffs the violets into a glass of water and sets the glass on the window ledge. He sits by the bed. Before long, he dozes. In the chair, he seems a pair of overalls with nobody inside.

Before, when she was still at home, when he was able to care for her there, she was awakened each night and many times each night by a housewifely summons heard through her dreams—a screen door to be closed or a cake to be taken out of the oven. He had always to be on guard then. Once, she had burned her hand with matches. Once, she had wandered out into the street. Then one day he pushed her bed against the wall. Thereafter he slept on the floor next to her bed. It was her bare feet stepping on his body that would awaken him.

"Come along now," he would murmur. "You don't want to be getting up. It's still dark." And he would nudge the woman back upon the bed.

"But the cake," she would protest, "it needs to come out of the oven."

"I'll take the cake out. You rest," he'd say, and lie back down on the floor, holding himself where her soft cool feet had pressed his belly.

Now, in the hospital, when he awakens, it is to the sound of water and gas churning in a pipe. It is an old sound to him, a sound equipped with fingers that reach for the deepest, cleanest place within him. It is a minute before he realizes that it is the sound of her breathing. In the mirror over the sink he catches sight of himself. The color of his eyes has faded; once, they had been blue. His head droops. He cannot hold it up. On the window ledge, the violets blaze on.

ROOM TWO

Melanoma. An eggplant and three plums plucked from the belly of a man. Split open upon a tray, the black meat bulged. Nor could the sprung flesh be restuffed in its casing; it was already too large. It is a wild pigment that races like bad news from the founding mole into the farthest reaches of the body. Listen! The sound of chewing. He is like an oak tree infested with gypsy moths.

Out in the solarium his wife and two young sons wait. The woman wears garnets in her earlobes and at her neck. Her eyebrows are darkly penciled, her lashes elongated, caked with dye. She is dressed in black. As though the man's tumor has metastasized conjugally. I tell her the news.

"It's a great loss," she says. "My husband was a brilliant man."

"He is not dead yet," I protest.

"Does he know?" I nod. "How did he take it?"

"He showed no reaction. He asked no questions. That will come later." The woman shakes her head.

"His mother used to punish him for getting sick," she says. All at once, her eyes lake and the mascara slides in runnels toward her mouth. One of the sons, the younger, lowers his eyes to cover the nakedness of his father that she has uncovered with her words. A silence. Then:

"Is it growing fast?"

"Very fast." I think of the gypsy moths.

With a sudden movement the woman places her hand on my arm, and a faint smile blackens her lips as well. I watch her large buttocks grinding down the corridor toward his room. Garnets are wifely stones, I think, for the touch of blood about them.

ROOM THREE

Goiter. A tall blond woman with a long neck. From the front of this neck a mass the size of a lemon rises. She sits up in the bed, and offers the tumor with a simple gesture of her hands. A tiny gold locket in the shape of a heart rests upon the promontory. To either side, a chain burrows beneath the collar of her bed jacket. To each beat of the woman's heart, the little gold heart responds with a shudder of its own. It has a brassy, impacted malice. It is the tumor's heart, I think. Perhaps all I have to do is reach behind her and undo the chain.

"Let me see," I say. And I go for the clasp.

ROOM FOUR

Acromegaly. Her pituitary gland makes too much growth hormone. She is a great tree, but the trunk only, with none of the arboreal graces. See how she fills the bed with the pile and strew of her bones. The bed is too short for her. Overreaching the end, the immense feet of a granite Assyrian queen. Her head is unfinished—unequal halves opposed. One jaw shovels the air about it, the other is dented. Her nose deviates to follow a smell at its left. As though, years ago in the womb, she had been set aside for some more insistent errand. A smudged work in progress with His thumbprints still upon it. The slow head swivels to face us, the eyelids half drawn, the mouth hanging. All at once, she sees us, and *something* softens the landscape of that face. The crags and ledges shift, are not where, a moment ago, they were. She smiles, and is completed.

ROOM FIVE

All night the old man bays at what, to him, I suppose, is the moon of dread but to me is the lamp on his bedstand. Each breath is a bone he attempts to dislodge from his throat. Such suffering, I think, must atone for any amount of wickedness. This man will die out of debt. Into the sheet each night he looses his saliva, urine, sweat and blood, impregnating the linen with his own oils and gums and resins until the sheets grow heavy while he grows ever lighter. Every morning a nurse comes to change the mummy's bed. But the old man does not die. Day after day, and many times each day, he must be turned in his bed, and lifted, and fed and injected. He has lost the control of his bowels and bladder. Pus drains from openings on his abdomen and thigh.

His wife is small and frail. She, too, is old.

"I'm taking him home," the wife tells me.

"But it will be too much for you," I say. "He needs too much tending to. A nursing home would be better . . . for . . . domiciliary care." She smiles at the word.

"His domicile is at 1834 Maple Street. That's our house."

"But what about all the mess?"

"How much mess can one man make?" she answers. And I am silent. For she has told me what I already knew, but had forgotten. Tending the body of another is an act of infinite loneliness, and carried out alone, a solitary commitment for which one is equipped only from the storehouse of his own heart. "How much mess can one man make?" I hear her words and I feel that I have been handed a secret letter which one day I must hand on to another, knowing all the while that this letter contains whatever there is to know about the care of the sick. When all else fails, the old woman teaches, take up reverence and proceed. In the throes of anguish one is far more likely to uncover his nobility than one is to show his cowardice. We are cowardly only in the expectation of distress. Once engaged, we shuck the ballast and, disencumbered, climb to higher ground. Disease magnifies both the sufferer and those who tend him. The patient in full dress of a wasting disease has no more clothing than a beggar. His skin is no more to be coveted than mendicant rags. His jaundice is a mark of autumn. Soon, he knows, it will be winter—cold and pulseless. No doctor worthy of the title will walk past such a dignified sadness unmoved, unstirred, to give of himself whatever might be of comfort. As one can love a beggar about whom one knows nothing, so can one love a dying patient who has openly declared the bankruptcy of his flesh. I have known a vain and selfish woman who had done no secular act more arduous then the application of cosmetics to her own face, yet who, when called upon, irrigated the colostomy of her dying husband, and dressed his oozing bedsores. Why deny even to Narcissus that the face in the pool may be that of another?—

Homer understood the power of the wound over those who both suffer and tend it. Toward the end of the *Odyssey* we find Ulysses at last deposited upon the shores of his homeland, the kingdom that he had left twenty years before. He is the very picture of the vagabond, dressed in rags and with his hair and beard tangled. Trembling with emotion he makes his way to the gates of his palace and gazes within. What he sees fills him with horror and dismay. The suitors of his wife, Penelope, have assumed control of the palace. They are conducting orgies and revelry; they are squandering his treasure, defiling his house. Penelope and their son, Telemachus, have been helpless to stop them.

Now the crafty Ulysses steps inside. He has decided to keep his own counsel, to let the tatters and dishevelment of twenty years of wandering hide his identity. He is led to Penelope, who engages him in conversation. She likes him, but she does not know him. She requests of him any news of her husband. He offers none. Sighing, Penelope calls for the old nurse, Eurycleia, to bathe the feet of the stranger and to make him welcome.

Now it had happened that forty years before, Ulysses as a boy of ten had gone to visit his grandfather in another country. While there, he had been taken on a boar hunt, during the course of which he had been gored in the thigh by the tusk of a wild boar. The wound was severe and quickly became septic. For a long time it was feared that the boy would die. It was this same nurse, this Eurycleia, who had then been placed in sole charge of the wounded Ulysses. She it was who bathed and treated the wound. I see her applying unguents and herbs, and pressing out the poisons, and cherishing the blood of her little patient as though it were the last of the wine. Until, at last, she saw that the necrotic tissue at the base of the wound had separated and sloughed. New pink buds of granulation tissue appeared. These coalesced to form a healthy bed across which the epithelium raced, and the wound healed.

It is forty years later. The old nurse kneels at the feet of the stranger. Suddenly she feels the scar on the man's thigh.

> This was the scar the old nurse recognized;
> She traced it under her spread hands, then let go,
> and into the basin fell the lower leg
> making the bronze clang, sloshing the water out.
> Then joy and anguish seized her heart; her eyes
> filled up with tears; her throat closed, and she whispered
> with hands held out to touch his chin:

"Oh yes! You are Ulysses! Ah, dear child!
I could not
See you until now—Not till I knew
My master's very body with my hands!"

It was not his wife or his son who had recognized Ulysses. It was his nurse. It was the wound that had awakened the buried past, the wound that was the emblem of all the shared pain and despair, the disappointment and the exhilaration that are the measure of the tending relationship.

WARD THREE NORTH

The General Surgical Service, Men's Ward. Rounds with Ora Guilfoyle, the Head Nurse. She is the kind of nurse who, by the mere smoothing of a pillow, can induce sleep in a febrile insomniac. We make Rounds in the manner of people who have come to count on each other over the years. We advance through the ward behind a dressing cart. At each bed we clatter to a halt. The chart is reviewed, the data recited. A wound is exposed, examined, redressed. We move on to the next bed. As we come to the end of the row, I am called to the telephone at the nurses' station at the other end of the hall. Ora is annoyed.

"I'll make it fast," I tell her. From the phone, I see her approach the next patient. She takes out her bandage scissors as though to begin removing the dressing from the man's leg. All at once I see her move to the head of the bed. She bends to peer into the face of the man lying there. Suddenly, she flings herself upon his body. One knee on the bed, and she is aboard, her skirt hiked. Now she straddles the man and bends to clamp his mouth with her own. As though her tongue were a key that would unlock the secret that lay in his body if only she could find the right way to insert it. She beats his chest with her fists, and huffs, blowing into a grate to keep a meager ember alive. The whole bed rattles and slides.

Such a passion would raise the dead. And so it did. Almost at once the man groans. A breath is taken. Another. Ora straightens, lifts her bruised purple lips away, pressing her mouth with the back of her hand, daring him to abandon her again. A minute later, Ora Guilfoyle has been replaced by the machinery of resuscitation. There is a team of doctors and nurses about the bed. They are all very young. A hectic gaiety prevails, a monitor beeps, a tube emerges from the man's windpipe.

Ora and I resume our Rounds. I am suddenly shy, silent. I think to say something that will acknowledge this event. But I do not. I have seen this woman at her fiercest—wild and desperate. I have seen the rhythmic jounce of dead men's feet. It is best to keep silent. We finish our work and wheel the cart to the nurses' station. A woman is there. It is the man's wife. From the distance, she has watched the coupling of her husband with this nurse. The woman raises one hand as if to speak to Ora. Ora hesitates. But the woman, too, does not speak. There is a glance between them. Then they move apart, the one toward, and the other away from, the bed where it took place.

MATERNITY WARD

The door to one of the Labor rooms is ajar. A young woman half sits up in the bed. She is moaning in pain. Another young woman leans at the bedside. With one hand she presses the back of the one in labor. With the other hand she rubs the huge belly, all the while murmuring. I know them! They are famous in New Haven. A year ago I had seen them fighting in the street. It was a summer evening on Congress Avenue where I had gone to make a house call. I emerged from my patient's house to see them come percolating out of a doorway. They had all the arrogant intentions of cooch dancers: a tangle of shrieks and fury, punching, slapping, clawing. One was fat and with a furious fat-man's face. The other was slim and mean-eyed. She had small translucent teeth. Both were blond, the big girl's hair close-cropped, that of the smaller one, ivory, to the waist. It tossed and writhed about its mistress, falling across the face of her opponent so that she could not see, wrapping itself about the big girl's neck, loyally fighting alongside its lifelong benefactor. Or was this hair an agitated wraith trying to make peace? The big girl swore from her throat. Her lips and teeth took no part in the words that barked up from her windpipe in solid lumps. "Fuckin' whore," she yelled. She was powerful and soon had the upper hand, beating the thin girl with relishment. But the desperation of the underdog was great and she found a way. Her translucent teeth indented an assertive, roaring breast.

Oblivious, the women had given birth to a circle of men who enclosed and nourished the struggle as though it were a flame that must be kept alive in a stiff wind. The faces of the men were wet with lust. They warmed themselves in the heat of the women. Their hands were in their pockets. They shifted from one leg to the other, whimpering and squirming like boys who need to urinate. I thought then of the ovaries and uteri of these two, and felt sick. What an elderly refined stomach I have, used only to the cooler colloquy of minds. But all that punching and blood, all that "fuck" in the air. There was a real possibility that I might vomit. I must leave this place, I thought. But I did not. I, too, had to watch. It was like spying on the faces of women in orgasm while they knew nothing but their own passion.

"You'd never know they were sisters," a man said.

Sisters! Of course! Only sisters could hate like this. Mere friends would never care so much. At last I could not stay. Blind and deaf I hurried to my car and drove away, only then daring to think of those two, how it was their way to cling and grapple and bite. How eagerly they accepted the tyranny of the flesh that had always, from infancy on, been offered and taken in love and sincere battle. The next day or the day after that, these sisters would slip the straps of their clothing to show the marks left by nails and teeth. Then they would laugh together heartily. One day they would help each other through husbandless childbirth.

THE EMERGENCY ROOM

A man has been stabbed in the neck. I undress the wound. A clot comes loose from his carotid artery and I hear his leaping blood cry out: Tallahassee! Tallahassee! Tallahassee! before I stifle it with pressure. When blood escapes from a rent in a large artery, each jet makes a little noise, not much more than the whisper of gas that escapes from a burning log. It has, I suppose, to do with eddying and the flow of liquid through a narrow place to the outside air. I don't know about that. But when the man died during the night, I wasn't as surprised as you might think to find out where he was born.

"Oh, come on!" you say. "You *must* have known it beforehand."

No matter. I know now that blood is a loyal conductor. It announces a man's native land just in time for him to get off the train at the place he holds nearest to his heart.

MINOR SURGERY

A thin, dark woman lies on the table. She looks to be forty years old. She is dressed in red, slacks and a blouse. Her eyelids are smeared with blue and silver; the lashes are long and caked. Near the outer corner of her left eye there is a small, star-shaped laceration. A bruise surrounds it. A few sutures will be needed to close the wound.

"How did it happen?" I ask her. She makes a small gesture of impatience. I see that it has been caused by something blunt. A fist, perhaps?

"Do a good job," she says. "My eyes are my only good feature."

"That's not true," I say. "You're very pretty." The skin of her face is thick with makeup. It comes off brown and grimy on my alcohol sponges. I must use many in order to cleanse the wound.

"Will I have a scar? My eyes are my only good feature." She says it again. Her hair is anthracite piled artfully. It is sticky to the touch and dry as hay. Not a strand moves when she turns her head to watch me draw the local anesthetic into a syringe. I must lift the hair away from her temple in order to work. When I do, I see the infected cyst just in front of her ear. A drop of pus hangs.

"What's that?" I ask. She makes no move as I inject. It is not pain that concerns her.

"I have an awful habit. I pick at it. There's one on the other side, too. I can't help it. I pick at them. I'm so nervous." I begin to stitch and tie.

"Can't you give me something to keep me from picking at myself? I'm so nervous. Valium helps me." Her mother stands nearby. She is a short, stolid woman also extravagantly made up and coiffed. She has an expressionless face.

"Please help her, Doctor," says the older woman.

"Give me thirty. Five milligrams," says the woman on the table.

"No," I say, "I couldn't do that. I'm a surgeon. I don't prescribe those drugs. Have you seen your family doctor lately?"

"He died," says the mother quickly.

"Give me forty. Just to keep me from picking. I'm such a nervous girl."

"Help her, Doctor. Give her something." They are like gypsies. Their voices sway together, chant.

"Please help her, Doctor. Help my daughter."

"Give me fifty. Fifty Valiums."

"You ought to do something about those infections," I tell her. "I will prescribe an antibiotic ointment. Apply warm compresses. When the drainage stops they should be removed. It is a smallish operation."

"Valium," she sings.

"Valium," they sing together.

"Help her, Doctor."

"Valium."

ROOM SIX

There are two sisters. They are neither young nor old. One, having married well, is rich. The other is poor. The rich sister has far-advanced leukemia. She will die unless she receives a bone marrow transplant. It is her only hope. The doctor of the rich sister has just asked the poor one to submit to an aspiration of her bone marrow in order to match the tissue with that of the patient. Only in that way, he tells her, can he predict the success or failure of the transplant.

"There is a good chance that it will work," he says. "With sisters." They are sitting together in his office.

"How do you do it?" asks the woman.

"A local anesthetic is injected in the skin over the breast-bone," he explains. "A tiny incision is made. Then a thick needle with a cutting edge is introduced. A small hole is bored in the crust of the bone and the marrow is sucked out. It is examined under the microscope."

"Does it hurt?"

"There is some pain. I cannot say there is not. Of course, we will try to keep it at a minimum."

"When does it hurt?"

"When the bone is punctured and when the marrow is drawn out."

Suddenly the woman shudders, then hugs herself as though her skeleton were a suitcase she was fearful of having stolen.

"No," she says, "I can't do it. I don't want to. I couldn't stand it."

"It is not really so bad. Perhaps I have made it sound worse than it is. It takes only half an hour."

"But then if I match up with her, you'll take a lot more."

"Yes. That is the hope."

"No. I can't. I won't do it. I'm sorry, don't ask me to. It is too much to ask—to give away my organs . . . my body like that. There's something wrong about it; it's not natural."

"But you have donated blood?"

"That's different. This is deeper. You drill a hole. It's tissue. You said so yourself."

"So is blood. Tissue. And you will make it up in a little while."

"Look, I'm not going to do it. You asked me, and I told you. The answer is no." She rises, takes out a handkerchief and dabs at her eyes, the corners of her mouth. "I'm a very nervous person." She begins to cry. "The idea of it—being sucked out. Like a boiled beef bone." The doctor is silent, but the woman knows that he is disappointed.

"What will you tell her? Does she know you are asking me?"

"I will tell her that you do not feel that you can do it. I will try to make her understand."

"Look, Doctor, I may as well tell you. We never got along. Even as children. She's never lifted a finger to help me. She's been lucky up till now. I'm the one that's had to struggle." Her voice rises with unexpected ferocity. "Did she give a damn about me? No! I don't owe her a thing. And I'm not going to do it."

"Then it is not because you are frightened?"

The woman starts as though caught off guard.

"Of course it is because I'm scared. The other has nothing to do with it."

"It's all right," says the doctor. He shows her to the door.

Now he is alone in his office. Cowardice, he thinks, is easier to admit to than hatred. It is also more acceptable.

I no sooner step out into the parking lot than I hear from high above plum-colored clouds the honking of geese. The only light in the sky is a moon so thinly carved as to have been a skin graft taken by a plastic surgeon. The geese cannot be seen. Their sound is all. I pause before getting into my car, listening. Do they honk to cheer each other up? Or is it a delirium that is the only state in which the long migration can be made? I am told they ride the air currents that stream back from each other's wings. But I think it is the cloud of honks they ride. On and on they flap, each adding to the sound that is the cushion that carries them, and that they must replenish.

Even so, not all of them make it. Now and then one goose will lose the rhythm of the chant and slip off, his tiny heart all honked out. Down he slews, delivering his small parcel to the ground. Far away and high above, for a little while there will be a gap. Then the others will shift a bit, drawing forward to fill in the line. Ever fainter the fallen goose hears the honking of his tribe until it is only the memory of a sound. Nothing to do with him, and he turns to face his lonely feral exile.

I think I need a vacation. If it weren't such a long trip, I'd go south for a while.

I am called back from home to the Emergency Room. An elderly woman lies on a stretcher. She is stout and gray-haired. Her apron is heavy and wet with blood. An hour ago, she had been beaten and robbed in her own kitchen. I bend to examine her smashed purple face. There is a smell about her, a wild smell. It is in her hair, upon her clothing. *His* smell. As though a wolf had urinated upon her before loping off.

"What did he look like?" the police ask. "Can you describe him?"

"It must have been awful for you, dear," says a nurse.

"He was afraid," says the old woman. "He shook all over as though to take a fit. I could see way into his eyes. After a while, when he didn't stop hitting me, I just stood there and hoped he'd kill me. I wanted him to. There wasn't another person in the whole world. Just him and me. He was the only one who could help me. But," she sighs, "he didn't."

ROOM SEVEN

The man does not yet know that he is dying. His wife sits in the corner. They are conversing.

"How do you feel, Frank?"

"I don't know. I'd have to read my chart to find out."

"What were you thinking about just now?"

"How embarrassing it all is."

"There's nothing embarrassing about it, Frank."

A small woman, she is made even tinier by the equipment that crowds the room—IV poles, cardiac monitor, cooling blanket, respiratory machine, suction.

"He says he has to pee and can't," she announces. I irrigate the indwelling catheter in his bladder. It is plugged. It will have to be changed. I send for the catheterization tray.

"Why don't you step outside for a few minutes, Edna? I'll call you when I'm done."

She stands and picks her way through the machinery to the door. I notice again that she is dainty and very good-looking. I see also that she really does not want to leave, that she is surrendering him to me. When she has left I remove the old catheter, put on sterile gloves and take up his penis in my left hand. I bathe it with antiseptic solution. Then I insert the new catheter and connect the tubing to a plastic bag. He sighs. I go to the solarium to find her.

"We're in hot water, aren't we?" she says.

I nod.

"I feel so sorry for him," she says. "I understand these things. I've lived through it. Two years ago I had a brain tumor removed. I was supposed to be ninety-five percent blind. But look at me." She smiles. "But Frank is so innocent. He's never been sick in his life. Such a good, good man. What should I do? Tell me. Should I stay or go home? Should I go down and put another quarter in the parking meter?"

"Go home for a while," I say. "Come back at eight o'clock. I'll tell him in the meantime."

"Oh, God," she says, remembering. "My mother phoned this morning and told me not to worry. It'll be all right, she said, because of the raspberry bushes. She gave him some raspberry bushes to plant. She said he'd be just fine in the spring when they bore fruit. What about radiation?"

"No," I say. It's too late for that.

"How do you do it?" she asks. "I feel sorry for *you*. Never mind about me. I'm tough. But you, how are you?"

"How are you?" I reply.

"It's all that equipment," she says. "The tubes, the suction, and that ghastly bed that moves up and down and sideways and does everything but make itself. I've lost him, haven't I? To the equipment. I sit in that room and feel him recede further and further away from me. Already I wonder if I loved him so much as I thought. The man with all those tubes in him—I never knew *him*."

She stands abruptly.

"For God's sake, peel it all off. Pull it all out. I want to give him a kiss."

At eight o'clock I am waiting for her in the room. Frank and all of the equipment have been taken away. The room is vast, barren. Edna and I sit on opposite sides of the empty bed. Our voices echo. We are hollow-voiced survivors.

"I'm sorry," I say.

"You should be," she says. "You gypped me."

Grand Rounds

Last week's letter was so full of cant and preacherei that I have been making up for it ever since by sinning all over the place—which for me means two more ounces of whiskey, three or four extra cigarettes and seconds of asparagus with hollandaise sauce. To say nothing of the usual flow of indecorous thoughts.

I have never been headlong in the matter of gallantry. Thus, I have wasted less time, and used up less precious energy than you shall, if your last letter was at all prophetic.

As for your obsession with the student nurses, take heed. There is a certain species of Arabian goat that never drinks throughout its lifetime, preferring to manufacture water from elements within its own body. This goat is content to live in the most arid and desiccated of lands. Should a pool of water suddenly appear at the hooves of such a goat, he would turn aside from it with profound disinterest. An intern ought to emulate this hircine wisdom, and make his affection out of his own dreams. He has neither the time nor the energy for anything but labor and study. But to such self-sufficiency one must be born. Once having pleasured the throat with cool water, it is no longer possible to do without. There is then only the winning of peace through hard work and repentance.

So, you are preparing a case for presentation at Grand Rounds. Choose your patient carefully and rehearse as if you were Pavarotti singing *Tosca* at the Met. The Chief who will attempt to destroy you in public is Baron Scarpia. You and your patient are the star-crossed lovers. Only remember that every heart in that vast amphitheater will be beating in synchrony with yours. Save that of Baron Scarpia, of course. Now let us to the Albany Medical College of thirty years ago:

It was the eve of Grand Rounds. Ah, bitter chill it was. My patient was Sadie Leckowitz, a longtime and famous sufferer from Schimmelpfennig's disease, for which affliction she was enjoying her eighteenth hospitalization in as many years. Her arrival on the ward was like the annual visit to a provincial hotel of a dowager of immense wealth and style. In fact, Sadie Leckowitz was a happy pauper unencumbered by material possessions. Once ensconced in the best bed on the ward, Sadie was ready to "receive."

All day long, maids, porters, orderlies, nurses and doctors paid her obeisance. Schimmelpfennig's is a rare disease, and Sadie was irresistible. Sadie Leckowitz was a veteran of eighteen Grand Rounds. It was to be my first.

She was seventy-four years old at the time of this admission to the hospital. I was able to arrive at this figure by studying the hospital record, averaging out the sundry ages to which Sadie admitted over the years, and then applying the fibonacci principle. Sadie owned a tireless penchant for conviviality and a whore's generosity with her organs.

"Hey, boychick, you wanna feel my liver? Come, I'll show you a liver and a half." And scooting flat in the bed she would haul up her johnnie shirt and present her abdomen, all the while submitting to the novice's maladroit fingers with patience and good humor.

"Here, I'll take a deep breath. See? It helps. Oy! Not so rough, Schmendrick. Did you feel? How many fingerbreadths did you get?"

"Two."

"Two! Two!" Sadie was outraged. "Two is for the shikker down the hall with cirrhosis. I'm three and a half. Feel again, you dope, you." She would share her indignation with the patient in the next bed. "For Schimmelpfennig's he gives two." And her eyes would glitter. Sadie experienced all of her joy through pathology. And Heaven in its munificence had made her a walking textbook of the same.

"Take a look at my knee," said Sadie. "Schimmelpfennig would come back from the grave for such a knee."

Two days before the event I asked Sadie whether she would be willing to appear with me at Grand Rounds. She hooked my arm with one rheumatoid claw and fixed me with a reproachful shaft.

"Dots all the time you give a person to get ready for Grand Rounds? What about the hair, the gown, the shtinkvasser? You men!"

For weeks I had read everything ever printed about Schimmelpfennig's disease. For these same weeks I had pored over the eight volumes of Sadie's hospital chart. Balzac, eat your heart out. Here is a *Comédie Humaine* that ought to be bound in tooled leather and edged with gold. When the banns were posted, I learned that my interrogator was to be the dreaded Professor Alistair Allanbrook. Six and a half feet tall, he was an oak among saplings. A single glance from that majesty reduced one to underbrush. Dr. Allanbrook was an awesomely well-informed clinician. And he had the kind of Anglo-Saxon severity that comes from having had no ancestral truck with the Mediterranean Sea.

Grand Rounds were held in an ancient amphitheater whose tiers of seats ascended at an alarming slant to within palpable range of the domed ceiling. The drama was played out in a circular stage at floor level. This was called The Pit. From The Pit the uppermost rows of seats appeared to lean inward until they were almost directly overhead. Grand Rounds were held every Saturday morning at eleven o'clock. They were attended on command by the entire faculty and student body, as well as by scores of visitors. I cannot remember when the place was not packed. More than one of my classmates had already foundered on these shoals, and I, oh, I, must now contend with Professor Allanbrook as well. The night before Grand Rounds I paid a last visit to Sadie Leckowitz in order to give her some preview of the questions that I would ask her during the opening event, which was the Taking of the History.

"It's bad luck to see the bride before the wedding," she said.

"I'm going to ask you a lot of questions tomorrow," I said.

"No kidding!"

"You just relax, Sadie. There's nothing to worry about. Everyone there wants to learn about Schimmelpfennig's disease. Please don't take anything that is said personally. And whatever you do, don't give away the diagnosis. It's supposed to be a kind of guessing game for the audience."

"You should live a hundred and twenty years," said Sadie.

"Well, get a good night's sleep," I told her. "And I'll pick you up in the morning."

"Break a leg, boychick." On the way out of her room I tripped over a footstool.

When I arrived on the ward in the morning, Sadie was already lying on a stretcher gabbing with an orderly. At sight of me she stopped in midsentence.

"Excuse me, dolling," she said to the orderly, "here's my date right on time."

"You look nice, Sadie," said a maid. "You goin' out on the town?"

Pushing the stretcher I saw the two yellow ribbons that had been tied into her hair. They had the look of butterflies hooked on an abandoned bird's nest. The stretcher was one of those diabolical gurneys whose front wheels were perpetually locked, the ball bearings having become impacted with a half century of hair, dried blood and God knows what lumps of detritus. Nor could these wheels be kicked free by the most vicious of boots. Steering was done by a combination of extreme body English and brute force, and even then, the stretcher followed its own instincts. At the big double doors to the amphitheater, I paused and peeked in. Oh, God! The place was full and waiting. The damned stretcher had made me late. Dr. Allanbrook was already enthroned. He nodded imperiously through the crack in the double doors, indicating, I supposed, his mounting thirst for my blood. Why am I so dumb? I thought. Why am I so ugly? So ill-fated? Why don't I have a deep, resonant voice?

I made several rather pathetic attempts to steer Sadie and the stretcher through the doorway, but the mule, sharing my reluctance to enter, bucked in a thirty-degree arc and skidded its wheels on the marble floor. In the end, I simply picked up the stretcher and its cargo and carried them into The Pit. One does not know one's strength until an emergency arises. Out of the corner of my eye I could see Sadie's lips moving soundlessly. She was counting the house. I took my place beside the stretcher. Dr. Allanbrook signaled me coldly to begin.

"Good morning, Mrs. Leckowitz," I said, pretending that I had not just carried her into The Pit in full view of the throng.

"You, too, I'm sure," she replied. She was playing her role to the hilt. I was proud of her.

"Now I'm going to ask you a few questions about your illness so that these doctors can get an idea of what you have been through." Sadie rolled her eyes in mock surprise.

"Mrs. Leckowitz, what brought you to the hospital?"

"I took a taxi," she said. A muffled sob rolled across the topmost tiers where my classmates clung to the railing like orangutans. A glance showed some of them with their heads buried in their arms. A few were biting hard on their textbooks. One at least seemed to be weeping, or vomiting. Dr. Allanbrook's face tightened a notch. Things were not going too well.

"I mean," I began again, "Mrs. Leckowitz, let me rephrase

that question. What was the matter with you that you had to take a taxi to the hospital?"

O treasonous tongue to have thus undone me! I had a feeling of impending doom. Perhaps I would have a pulmonary embolus.

"Never mind!" I said quickly. "Don't answer that." And I stood ready to clap my hand over her mouth if necessary. Too late. With a kind of abstract horror I heard Sadie say:

"Why I had to take a taxi? Because my horse got Schimmelpfennig's. He caught it from me."

If my years be Methuselan I shall never recover from the sight of the mass epileptiform seizure that swept the amphitheater. Faculty, students, visitors, all were caught up in it. What, in God's name, could it mean? I closed my eyes. For a moment I was standing in a medieval town. It is five hundred years ago. The houses on either side rise and lean over the street. Almost, they touch. An instant too late, I hear the cry: *"Gardez l'eau!"* And I look up to receive full in the face the pailful of slops that has been tossed. I wipe my eyes and see from each high window the strangled face of one of my friends.

In the confusion that followed, I whispered to Sadie out of the corner of my mouth.

"Impostrix," I hissed, "you gave it away," and sank six inches deeper into my shoes. I stood by the stretcher, examining the floor for some sign, while Sadie, having captured the crowd, now encouraged them with little moues and wavings. All at once, a pair of shoes stepped into my downcast field of vision. Large, oxblood, wing-tip, sadomasochistic shoes. I heard the frenzy begin to subside. Even Sadie looked up expectantly. When I lifted my gaze, it was to see Professor Alistair Allanbrook standing next to me. He spoke:

"I think we can let Mrs. Leckowitz go back to her ward now," he said in the baritone of Ezio Pinza. And turning to Sadie, he bowed with all the grace of a courtier at the palace of Louis XIV. "Thank you so much for being with us this morning, Mrs. Leckowitz. You have, as usual, been a great help."

"My pleasure," said Sadie.

Once again the stretcher and I did battle, with Sadie waving good-bye to everyone over my exertions. At last she was outside the door.

"Wait here," I said coldly. "Someone will come to take you back in a few minutes."

"You were terrific," said Sadie. "A knockout." But I had already turned back toward The Pit and my martyrdom.

A Pint of Blood

When I was, as you are now, an intern in Surgery, I was a man of unbounded optimism. Mine was a cloudless sky. Love, of which it is often said that there ain't enough to go around, was, if anything, in surfeit. I was mindlessly content to be where and what I was. It was in the midst of my rotation in the Emergency Room that I received a summons from the Chief of Surgery, a dour thoracic surgeon of truly Arctic conviviality, in whose capable hands had lain the lungs of ten thousand human beings. These same hands now held the professional fate of thirteen shabby, exhausted interns, twelve of whom swung from mania to melancholia with a speed that was dazzling. I alone remained unfailingly confident of the future. Everything would be all right, I counseled my colleagues. Not to worry. We would all be invited to return the following year to begin our Residencies. This, despite the inescapable fact that for thirteen interns there were only six openings at the First-Year Residency level. Seven of us would be dismissed. But I could not think of that.

Within minutes after the summons from the Chief of Surgery, the news had spread throughout the length and breadth of the Emergency Room. Be still, my heart, I cautioned myself. I knew at once that I had been chosen for one of the coveted positions. On this very day, I was to hear the announcement. Benjamin Franklin was right, I thought: Prudence, frugality, industry—it all pays off. More hearts than mine beat wildly at the prospect of my interview with the Chief, some, yes, with envy, some with admiration, and one, at least, with pride. The Head Nurse in the Emergency Room was a young woman named Rosalind. Eyeing me with dispassion, Rosalind suggested that I rush to my room in the Interns' Quarters and change into a clean set of "whites." When I looked down at my uniform, her meaning became clear. My "whites" were nonesuch, but instead were a blend of red, black, yellow and green, a wild canvas depicting the dark side of metropolitan life. "Furthermore, you stink," said Rosalind. And it was true, the odor being a mixture of sweat, vomitus and something goaty.

I raced to my room and peeled off my foul rags. I washed my face, combed my hair and forced myself into the savagely starched sleeves and legs of my only clean uniform. Thus shining in lilied raiment, I presented myself before the Great Desk in the Office of the Chief of Surgery. Behind this desk he sat, twirling a pencil between the fingers of both hands in the manner of powerful men everywhere. It was less an official act than a mysterious ritual to awe and rebuke the lowly. Never once did he glance up to acknowledge my presence. I waited, thinking idly that his heavy jowls and white sideburns were a fitting contrast to my own coltish spank. At last he spoke.

"It has come to my attention, Doctor, that you have lost a pint of blood."

"Sir!"

"You have, it is said, lost a pint of blood. Do you deny it?" O Heaven! O Fate! What dark business here impends?

"Why, sir, I have sustained no injury to my person that would result in the smallest shedding of my blood. Unless you consider an insignificant chin wound suffered in the course of shaving. . . ."

"Then you do not deny it?" The tone of his voice informed me that I must above all be agreeable.

"No, sir. If you say that I have lost a pint of blood, why, then, of course I did. No, sir, I definitely do not deny it. In fact, I would, if you wish, vouch for it."

"So!" The Chief swiveled his chair so that he no longer faced me, an arrangement that made it possible for me to speak again.

"May I beg your indulgence, sir, that you might tell me when and how the news of my hemorrhage reached your ears?"

In cadence grim and Hiawathan the Chief recounted how, in the dead of a night punctuated by the horrors of mortal affliction, I had gone to the Blood Bank and signed out two pints of blood which I meant to administer to a man who was even then bleeding copiously from his duodenal ulcer. One of these pints of blood was duly hung upon the IV pole and run in at full speed. It had just barely been absorbed when the patient sat up in bed and, with a single cough, produced his entire blood volume into the sheets. Within moments the poor man had gone to his reward. In the press that ensued, what with the heroic efforts to resuscitate the patient, all of which were of no avail, the second pint of blood was left sitting near the radiator in his room for a period of time sufficient to render it useless to any other patient. A greasy dawn had just slid through the windows of the ward when this fact was discovered. That second bottle of blood had become part of the wreckage of that ill-fated night. Such was the tale recounted for me (in somewhat different terms) by the Chief of Surgery. When he had finished, I stood, leaning against the starch of my uniform, listening to the thutter of snare drums. As the white bird of hope flew out the window, I prepared myself for the ceremony in which he would snip the tubing of my stethoscope. Once again he spoke those terrible words.

"You have lost a pint of blood."

"Sir."

"You will replace it with a pint of your own at once."

"Sir."

So ended my interview with the Chief of Surgery. I who had ten minutes before raced from the Emergency Room in *exsultate jubilate* crept back *in requiem*.

How to describe the emotions of my friends and colleagues upon my return to the Emergency Room? I saw their faces cloud up with sorrow as I told them what had transpired. Even in the abyss of my despond, my heart went out to them.

"That bastard!" said Rosalind.

"That son of a bitch!" said my fellow intern.

And so it went for half an hour. Evidences of compassion were everywhere. All at once, Rosalind, she of the green eyes and vivid tongue, spoke up:

"You can't spare a pint of blood. It would be murder. Why, you have hardly enough to get by on yourself." Murmurs of assent, much nodding of heads.

"Just so," said Rosalind, warming to her subject, "just so does the wolf seek out the weakest, the most enfeebled member of the herd."

By now, I was wildly in love with her. How beautiful she was at that moment. Her sea-green eyes flashed. Perhaps, later, she might console me properly as I lay dying of anemia. I removed my white coat with difficulty (the starch), and rolled up one shirt-sleeve. Every face winced and turned aside.

"Jesus," said my colleague, "if it looks as though you are going, we'll raise the bottle a little and run it back in." I pressed his shoulder, knowing full well what that would mean to his own chances for a Residency. I lay down upon the stretcher, held out my arm and requested that he, my fellow intern, be the one to draw the blood.

"Me? Draw?" He was clearly horrified, as though I were Brutus and he the reluctant Volumnius on the Plains of Philippi. Thus were we frozen in a kind of martial tableau, and would doubtless have remained so to this day had not a deep growling voice spoken softly at that moment.

"Somebody need blood?" It was Clyde, the orderly. Clyde, bursting with the juices of life. Clyde, the Saint. As one, we stared at him in wild surmise.

"I got blood I ain't even used yet," said Clyde, the Holy Man. In a trice it was Clyde upon the stretcher. Another trice, and his blood was in the bottle. The third trice had me back in the Office of the Chief with the bottle of Clyde's blood in my hand. Its lovely warmth remains among the most memorable pleasures this hand of mine has ever enjoyed. I cleared my throat.

"Here you are, sir. Fresh from the basilic vein. And still warm, as you may prefer to confirm?" I held the bottle close for him to feel.

"Take it to the Blood Bank, and sign it in," said the Chief.

I had already turned to leave when the Chief called out my actual name! "Oh, by the way," he said, "you will be coming back to us next year for your Residency. Congratulations."

Letter to a Young Surgeon
IV

———————————◆·◆·◆———————————

Let us go. It is seven-thirty in the morning. The men's locker room is a scene of great activity. Surgeons, Residents, Interns and Students are dressing for the morning's work. There is a nervousness in the stalls, a nickering and stamping. These surgeons are neat; they hang their clothing in the lockers with precision; everything is folded along its creases. Paper boots are fitted over shoes with care, caps tied behind the head just so, the strings of the masks knotted and adjusted. Metal doors clang. The bodies of these men are like the furniture of your house; you have seen them every day for years. They seem beautiful and young, even the old ones. Their hands are pink and warm to the touch from so many scrubbings. Each fingernail is trimmed in a gentle curve. Their talk is all of football and baseball and golf. Surgeons love the notion of teams pitted one against the other, or of single combat. It is fitting that this be so. In the corridors the patients wait. They lie narcotized, silent, yet afraid, on stretchers, each one parked outside the room into which he will be taken. As each surgeon approaches his patient, he, too, grows quiet. Who knows what dreams these surgeons have had?

The women surgeons dress with the nurses. Never mind. The inheritance of surgery is less a patrimony than a matrimony. "Male" is inspired tinkering, a knack for repairing the works. "Female" is softness of touch, an intuition for planes and membranes. Like the poet, the surgeon must incorporate both. If pressed I should confess to the suspicion that women make better surgeons than men. Women know how to fold themselves about life. Women know blood, and they know pain.

The light in the operating room is no less important than the light in an artist's studio. It must be direct, cast no shadows, yet be free of glare. None of these has ever, to my knowledge, been achieved to perfection. It is to be expected then that a surgeon will complain of the poor light forced upon him by malicious electrical influences. But inferior light must not become the excuse behind which a nervous surgeon conceals his fumbling. Operating rooms must be of a certain size. Too small, and there is more likelihood of contamination by the collision of occupants and equipment. Too large, and the mind is distracted. Great empty spaces take away from the proceedings that lovely feeling of a close-knit team on an expedition. Of the two, I prefer small rooms which act to concentrate the thought and discipline the mind. Too much space smothers me. Being slight, I am able to insinuate myself amongst the matériel. Bulkier members of the species are less insertible. Let them have wider ranges.

You wheel the stretcher alongside the operating table and help the patient to slide himself from one to the other. Now you must keep in contact with your patient until he is asleep. Hold his hand if one is free of the anesthetist's use, or place your hand upon his chest, his shoulder. This will suggest to him that a protective spirit hovers over him. Speak quietly, as though privately, to your patient. Dwell upon his awakening. Say the word "recovery." It is a winged word. The sight and sound of you at this time is balm in Gilead to him. No matter how kindly the others are, he does not know them. You are his doctor, the only one who has received his trust. These others are strangers. Tangency in the operating room is both given and received. Often the patient will reach out to touch you, taking comfort from the feel of your body, just as you draw strength from his. The patient is not a plant of the species *noli me tangere*, whose seed vessels would burst at the slightest touch. He is frightened, feels himself to be alone in a green and clanking place where there are no windows and he cannot see the sky. Let him look into your eyes for whatever distance and space he can find there.

One day you, too, will be a surgeon and stand over a table upon which a patient lies. In a few minutes you will take up a scalpel and lay open his body. Your patient may be, like the man I operated on today, old, wasted, his bowel obstructed by a tumor. All at once, before the induction of anesthesia, this man reached up his two hands. They trembled in the air like weightless dragonflies. The long papery fingers encircled my neck. His grip had a strength that surprised me. Not to strangle but to draw in through his hands all the blood and breath that coursed there. To this, you, too, will gladly submit, knowing that it is a greater offering than any mere surgery. It is only human love that keeps this from being the act of two madmen.

Take care not to get in the way of the anesthetist. During the induction of anesthesia, you have the chance to do small tasks that might assist him—tearing strips of adhesive tape for him to use in securing a needle in a vein, or pressing, if he should request it, the larynx of the patient to ease the passage of his tube into the trachea. It is a privilege to attend to the needs of an anesthetist.

If the anesthetic is to be spinal, help the patient to turn to one side and to curl his body. Now cradle and steady him until it is done. Never fear to be thought unmanly because of such service. Now your patient is ready to go to sleep. To wish each other luck is neither inappropriate nor insensitive. It is honest. The injection of Pentothal is given. You continue to soothe him with your touch and voice. Is there no hidden cabin of the brain, I have wondered, some dark sulcus where he will hold these words you say to him: "Relax. Go to sleep. Pick out a nice dream"?

"That's right, give in. Don't fight it," you say. And he submits. Perhaps one day he will summon these words forth to the blazing courtyard of consciousness. Then he will hear them again and think of you, and smile. Never doubt that he has heard you. Sounds are intensified during the induction of anesthesia. Now he is asleep and hears nothing. Anesthesia is an imitation of death, uninhabited even by dreams.

Like all ceremonies, washing the hands has a wisdom beyond mere practical worth. Like the donning of clean, loose-fitting clothing, it is an act of supplication. It goes without saying that the fingernails are kept short and straight and clean. There must be no sharp points to snag flesh. Nothing sets a patient's pulse to hammering like the sight of a cache of dirt beneath his surgeon's fingernails, for it implies carelessness with the flesh of others. (Between you and me, I can think of one or two surgeons whose natural corruption is such that an impaction of good clean New England dirt under the nails would be absolutely purifying. When I told this to my nurse, Anne Palkowski, she said to give her a minute and she could think of several more. Give her an *hour*, and the whole thing would become a national scandal.)

Take a pointed stick and wipe beneath each nail. Throw away the stick and take up the heavy-bristled brush. Drench it with soap and water. Proceed to scrub a finger at a time, then the thumb— front and back. Advance to the dorsum of the hand, firmly, and with ever-increasing circles. On to the palm. Flex the fingertips, back and forth, back and forth. Now in long, luxurious strokes, up and down the arm, to an inch above the elbow. Rinse, and do it again.

The body of the patient is now uncovered. In the matter of such exposure the surgeon must set the tone for the others to follow.

Remember that you and the sleeping patient are one and the same. To expose his nakedness is also to expose your own. Therefore draw upon yourself the mantle of modesty. There are occasions that require the genitalia of your patient to be shown, either for examination or surgery or for teaching. Accept such exposure with a dignified and natural mien. Should the patient be awake during this time, help him to bear whatever discomfort or embarrassment this engenders.

If nakedness is not essential to the proceedings, draw the sheets about the patient so as to cover what otherwise might go uncovered. To cover the nakedness of another is an act of charity. To gape needlessly is to be like the lewd elders who spied upon Susannah sleeping in her garden. This is also the way you must instruct the novice medical student. Embarrassed and threatened by being witness to such things, the student will cast about like a floundering swimmer for the reactions of the others present. He will gratefully anchor himself to your calm, undismayed face, and he will set his own in the same mold. The young woman being readied for appendectomy is not Aphrodite making an earthly appearance. Nor is the surgeon Praxiteles sculpting Venus. She *is* a work of art, but not by you created.

Doctors who indulge their lust among the patients get nothing but a bad reputation and a guilty conscience. They shall feed upon the dry cobs of repentance. In an act of passionate love the two participants are nude. In the love between patient and doctor they are naked. Both are vulnerable, laid open to chance and design. The true healer retains the nakedness of a child; he remains open to the emotion of love for his patient.

But let us be honest. To gaze upon the beautiful unclothed body is a spur to the lustful imagination. Soon, in your mind, you are fondling this body, kissing it. To deny the existence of these urges is to deny your humanity. To fail to suppress these urges is to accept a condition of bestiality. Suppress, therefore, and do not feel guilty for your lustful thoughts. Even the gods hankered. Only distinguish between nakedness and nudity. It will enable you one day to be touched to the heart by the vanished bosom of an old woman.

The surgeon takes incorrigible delight in the immersion of his own body in that of another. It is a kind of love. But it is never to be confused with eroticism. Dwelling as he does within his patient's body, when a surgeon makes an incision, it is a self-inflicted wound.

Poised above the patient the surgeon is like a priest guarding and preserving fire. He takes strength from this closeness. For the body of the patient is the sun, the whorl of light and heat that radiates life into this room. It is the patient's heat that foments this work, his light that makes it visible and possible. Without the patient this world is dead. He is the nucleus, all the rest, the cytoplasm. He, the seed; the rest, the fruit about him.

When the incision is made, the surgeon gazes through the aperture, and even as he does, something marvelous happens. He himself shrinks to accommodate to the dimensions of this unexplored place. Once, while studying a group of tiny bonsai trees in a tray, I knelt to peer upward into their branches from below. Suddenly it was I who was small and the trees that were large. I was a pygmy in a towering forest. Such is the magic of bonsai, and such is the magic of Surgery.

The abdomen at the same time expands to incorporate him. He descends, reliving his childhood fears. He is once again in a mad, dark cellar with, high above, the fading light of earth.

Once inside, every artery is a red river to be forded or dammed up; every organ, a mountain to be skirted or climbed. The surgeon is a trekker in a land of mystery and beauty. Beneath his feet he feels the throbbing of a great engine. About his head a warm wind breathes. No less than the one he has just left, this new world holds all there is of good and evil. An intimate warmth encloses him. It has the humidity of his mother's unremembered womb. Nor does he once look back at what recedes outside. From the moment of his entry he is totally engaged. He is in a state of *topophilia,* wherein his vision changes. He sees vast dripping caverns, escarpments, lofty ranges. Pink and salmon and maroon creatures drowse in their beds. These creatures are friendly; the surgeon approaches them with affection. But there are fumes, too, a sluggish trickle. Far in the distance a stony gray crater presents itself. One, and then another, and another for as far as the eye can see. There are the very plains of Hell.

For years the surgeon has dreamt of this place, conjured it in his mind, imagined all of its marvels. And now he is there. Unlike the poet's, whose travels are figments and phantasms, this is real.

At first it seems that sound, too, has shrunk. There are only vast silences. Soon these silences are violated by the whoosh of blood coursing through hidden passageways and the throbbing of an engine far below. The surgeon experiences systole and diastole, not his own, but a rhythm absorbed from the landscape. A distant wind blows to and fro, folding and unfolding the fine linen that hangs between the flanks. But wait awhile and there are other sounds— whispers, giggling, murmurs, ticktocks retrieved from beyond the auditory threshold.

Once having shucked the outside world, the surgeon's mind has been set free. He is once again a child gazing with the liberated eyes of childhood. But a wise child, for he brings here all the reason and logic and knowledge of that other life. Like a mountaineer who must bring along his own food and oxygen, the surgeon feeds upon his past experience. It sustains him throughout his journey. But he must be certain to bring enough. Otherwise he will surely die in this unterrestrial place. Who would visit this awful place must be like a deer, gifted with presentiment. He must stand as the deer stands —motionless, breathless, quivering, to feel the distant sounds come nearer, the sounds that no one else can hear. He lays his ear to a kidney to hear its soft watery machinery. He listens to the filtration and extraction of the liver, the muddy slide of the gut, the ceaseless chewing of a tumor. He hears the whole cavern of the belly reddening, as though he were under the influence of hashish.

For the surgeon, the distance between the lips of his incision and the prize he seeks, and for which he has been dispatched by his patron, is no more than twelve inches. Still there are gorges to be crossed on narrow swaying filaments, precipices to be scaled, Fate to be placated. All the while on his trek, the surgeon's mind urges toward that tumor. His eye is fixed upon its gray stony parapets. From the moment of incision he is already laying siege to it, tying off its supply routes, burning off the foliage that protects and conceals it, tightening his grip on it, even dwelling within it like a spy, gathering information, taking measurements, looking for signs of weakness. The surgeon hates and loves this tumor that both resists and offers itself to him. It is the object of his dreams. At last, he retrieves the tumor from this dream and holds it up to the light. Hunter and prey have met in one. A man among men in one world, he is a man alone and lonely in this other. Here within, he shares none of his terror or exhilaration. There is only the solitude of terrible works.

Now the expedition is completed. In jubilation or despair, the abdomen is closed. A line of sutures like the lashes of a closed eye marks the passageway. Already the scar begins to form across the entrance—blood, serum, a jelly of fibroblasts and tissue juice gather, all the cellular throng that will barricade with boulders and walls of flesh.

Safe outside, the surgeon turns to gaze at the abdomen, imagining still the maroon and humid walls, the bare gray crags, the moment of victory or defeat. And the pulse beats faster at his wrist. Now the wound is dressed. How quickly an incision becomes a wound! The unbandaged eyes of the surgeon see beyond the sealed gate to the garden within where once, like Adam, he walked in high discovery. This sealed and silent abdomen, this, too, will never be the same as it was. For the surgeon has left his mark hewn into a remote trunk. All about the tumor he has built an altar of metal clips arranged under auspices.

His fingerprint is on the wall. And he has left his dreams behind to glide among the viscera, mute witnesses to what has transpired. The abdomen is closed; the surgeon's dreams remain within. Nor are the organs of this abdomen ever wholly forgotten.

They have become part of the surgeon's past. And these organs have imparted to him a certain knowledge, whispered to him secrets that he will pass on to others. Much as a jewel box contains the dreams of a vain woman, or a casket of bones the dreams of a widow, inside this closed belly flit the fitful dreams of the surgeon.

Letter to a Young Surgeon

V

In the moments before surgery, look, if you will, at the tray upon which the instruments lie. There is a certain restlessness among the hemostats and forceps. Do you detect it? A cabin fever, like the eagerness of hounds wanting to be set free in the woods. Even surgeons who have never learned the knack of loving people will love these simple, hard, shiny tools, knowing that they will not betray him.

Surgical instruments, like the brush and palette knife of the artist, possess a mystic anonymity. It is only when they are taken into the hand and guided that they become identifiable as tools or implements. A knife is but a knife until it becomes a scalpel. Only then is it exalted. What a far cry these glittering tools from the fleams and stylets of ancient man, yet no more beautiful for all their complexity. The man who took up the splintered thighbone of a gelded lamb in order to couch the cataract of his tribes fellow was no less a surgeon than the man who squeezes the trigger of a laser gun.

We are too removed from our tools. It is a kind of betrayal of our craft. Because we do not make our instruments with our own hands, we do not use them to best advantage. Our relationship to our instruments is a more distant and formal one than were it still necessary to chisel the nib of a knife from a stone or sharpen a goose quill into a probe. A great oboist cuts and ties his own reeds. So does the fisherman his trout flies. In so doing these men infuse them with their spirit. If the gift of prophecy has not deserted me, I should say that such men will make sweeter music and catch bigger fish than those who place their craft in the hands of manufacturers. There is something soulless about a steel blade that is punched out identical to all others by a machine, is used one time and then discarded.

An ancient tradition commands that on the night prior to a single combat that is to decide the fate of an empire, or on the eve of one's coronation, one must perform certain rites so as to be in a condition of greatest strength and purity. To do so is but to show reverence for the task ahead. Cleansing of the spirit is achieved through prayer and meditation; cleansing of the body, through bathing, fasting and abstinence from sex. It does not seem to me overly harsh to require of a man who will be crowned king in the morning that he absent himself from palatine and penile felicity on the night before. One's coronation is not, after all, a Monday, Wednesday and Friday affair. But were we who practice surgery to so deprive ourselves on every operation eve, I very much fear that no patient would be safe from our blood lust!

It is at least as ancient a custom that a hero wear upon his person, at the moment of combat, some artifact, preferably small, and not cumbersome, a woman's glove, say, or a locket given him at the outset of a dangerous quest. Having once prevailed over great odds while carrying such an amulet, the hero comes to regard the piece as both dear and necessary. He could not any longer fight or brave the elements without it. He would do so at his own peril. Once inside an arena, having forgotten to bring along his favorite charm, a matador would be lucky indeed to depart the ring with his own ears and tail intact. Such recognition of the power of magic is not to be gainsaid. The neurosurgeon who pins the last tatters of his baby blanket to his scrub suit just prior to the performance of a craniotomy is neither to be pitied nor censured. He is obeying a time-honored command of the spirit. Such a man is to be trusted. Far better the fetishistic, amulet-loving surgeon who, while not devout perhaps, loves his charms. He is likely to love his scalpel and his colon clamp the way a Bedouin loves his camel.

An even older tradition insisted that the victor in combat devour the heart, brain or testicles of his vanquished foe. By this ingestion the winner took upon himself the courage, strength and sexual prowess of his enemy. I shall say no more about this save that such behavior is altogether inappropriate at a teaching hospital.

Along with the heroic tradition, much of the hero's love for weapons has disappeared. Some surgeons like to use only the newest gadgetry, with gleaming unscratched surfaces. I should much prefer the comfort of old instruments that I have used over and over for years. How lightly they would leap to hand, with joy, almost, like a horse taken from his stable and mounted for a ride. He is delighted to be made use of. It is why he was born. So it is with old tools.

Breathes there yet a surgeon who, faced with the prospect of an emergency splenectomy, calls for his lucky retractor? Who sings a serenade to his scissors, a hymn to his hemostats? Where is the orthopedist who would strum a lute in praise of one certain periosteal elevator? Where, at last, is the myth of the hero of this work, wherein a young intern of humble birth, making his first incision, is greeted by a mysterious hand emerging from the abdomen holding an elaborately carved, gem-encrusted scalpel? Trembling with pride and awe, the chosen intern receives the scalpel into his own hand. It is to be his for the rest of his life. With it he shall lay waste a mountain of tumors and ulcers and inflamed gallbladders. Commanded by its master to perform an appendectomy, Excalibur leaps to the task of its own accord, needing no further encouragement or restraint. Between operations, it is worn at the belt in a stoppered vial of alcohol. What with all of our knife blades, drapes, towels, cauteries, masks and gloves being of that transient type called throwaway, it is only the surgeon and his patient who are not discardable.

In the absence of any such magic scalpel, surgeons are expected to perform heroic deeds with a race of detachable knife blades that do not differ one from the other by so much as a jot. It is no wonder that Surgery is in public ill-odor when it is practiced by drudges who hold in their hands each day a different knife that is scarcely noticed even while it is being used, and is remembered not a second longer than it takes to make the single incision of its lifetime. I used to know a surgeon who, if handed an unwanted instrument by his hapless scrub-nurse, would hurl it across the room, where it would strike the wall and fall to the floor with a loud clatter. There was a time when this sort of tantrum was not only permissible, but in some absurd way, admirable. One smiled, as one smiles upon a clever brat.

But I, for one, depend too much upon my tools to throw them about. I *believe* in them. And *things* remember. The scalpel that shakes off its restraints and noses one millimeter too far into the tissues, cracking open the inferior vena cava—is it redressing an old grievance? Has it been smarting under some long-nurtured insult? An old mistake of yours or some other surgeon's? This is not to say that, out of some misguided notion of *politeness*, you should hesitate to demand a replacement for a scalpel or hemostat that seems to you at that moment less than perfect. It is of no importance that you be deemed crotchety, cantankerous or curmudgeonly. Which, of course, you are, you are. And which you have every right to be if you spend your whole life doing surgery. Who can blame a bear if now and then he rises up growling, makes a menacing pass with his claws, then settles back down with a huff and a thump and a cloud of dust? And so it is with surgeons.

I am not blind to the possibility that one day a knife will turn on him who holds it, will present its belly to that of the surgeon and demand that he, too, surrender.

Not long ago, I pricked my finger with the scalpel with which I had just lanced a boil containing bacteria that had proven resistant to all of the antibiotics. That night, my finger itched. I rubbed it gently, knowing what was gathering there, helpless to do anything about it. By noon of the next day there was pain and swelling. At midnight, my temperature was 103. Dread and fury set in. I wished for the owner of that boil to die. As though somehow that would even the score. I squeezed the skin around the wound, hoping to express a few drops of infected blood, much as a dog will lick and lick the wound it has received in battle with another dog.

It was of no use. With a grim precognition, I watched erythema and induration ascend—past wrist and into forearm. Ahead lay the strategic elbow. And along with the redness and hardness, there mounted a splendid pain. O narcotic nights! Full of graveyard decay . . . But I have survived. I know not how nor know I why. Only that, at the very doorposts, the lintel of the elbow, the red wave restrained itself, held at that place for one long day, then slowly fell back, disappearing into the finger from which it had escaped, then paling out until it was gone. No. A discardable, mass-produced knife blade has no history. Its past is as absent as its future. Give me hand-hewn quills and axes and chisels. And sutures twisted and braided by patient fingers and stiffened with wax, or made from the woven hair of Japanese monks. It is dismay at the fading from surgery of instrumental devotion that prompts me to set down for now and forever the instructions which, if followed to the letter, will enable you to make your own scalpel:

1. Arise before daybreak on the morning of the operation. After bathing and donning clean raiment, meditate for the period of one hour, washing your mind clean of shallow thoughts. Concentrate all this while upon the form and shape and size of the scalpel you must make. As though you would realize it out of thin air if you could.

2. Go alone into the depths of a forest, carrying only a shovel and a stone ax, permitting instinct to lead you to the proper site. You will know it at once. As like as not, it will be a small clearing among the trees so that the thickest roots will not present an obstacle. It is a place as bright as childhood.

3. Begin to dig, and continue to dig for a depth of one of your legs. Soon your shovel will strike the surface of a rock. If

this rock be cleared of earth and measure at least one hands-breadth, you have found what you have been seeking.

4. Carefully, so as not to crumble or gouge the rock, clear it of its encasement of earth. Now insert your fingers in the trough you have made all about it, and lift the rock out of the ground.

5. Place the rock upon a previously laid bed of dry leaves, preferably oak for majesty. Oak leaves are most likely to have dwelt on high and so will have retained the hint of loftiness.

6. Now take up the rock once again, holding it by one rounded surface only. With the stone ax, strike the rock a sharp blow such that it will be bisected cleanly. One half of the rock will fall to the ground. The cut surface of the piece you hold will be seen to be dry, smooth and utterly clean.

7. The second blow is the more difficult. It is understood that the novice will require any number of blows and will demolish many rocks before his work is blessed with success. The second blow is administered to the half that remains in your hand such that the force of the blow is not perpendicular to the cut surface, but, rather, angled toward it. A second portion of the rock is allowed to fall onto the dry leaves. The pie-shaped portion remaining in the hand will have a sharp terminal edge that, if the second blow has been angled correctly, will be comprised of what was formerly the central substance, the *gist* of the rock. Thus it will never have been previously exposed to the air. It goes without saying that the new fresh edge of what is now your scalpel must not be touched, nor must it be allowed to touch anything. Only the thicker "handle" is to come in contact with your hand.

8. Examine the cutting edge by holding it up to the light. If your blows have been sharp and quick enough, the edge will be a smooth, unchipped line. Do not test its sharpness with your finger, although the temptation to do so will be all but irresistible. Have faith!

9. The fatter "handle" must sit comfortably in your palm so that your index finger, when laid along the side of the knife, can guide and control the cutting edge.

10. Now take the knife as it is, in your hand, to the place where the patient waits at the edge of the forest. He has been carried there by his kinsmen. His wound is a great bristling abscess of the thigh where, like the young Ulysses, he has been gored by the tusk of a wild boar. The leg itself is hugely swollen, tense, shiny and fiery red. The skin of the leg is hot with fever. The man moans and cries out in delirium, and

grips his thigh above the wound as though to strangle a rat that gnaws him.

11. Kneeling by your patient (kneeling is the ideal posture for surgery as for prayer), murmur to him in a voice that is low and steady. Do not doubt that through his seizures he hears you and is comforted by your voice. Some patients will, at this point, grow quiet and hold still.

12. Steady the thigh just above the wound with your free hand. Now, with your stone make a sudden quick slash across the wound at the point where the swelling is greatest. The exact length of the slice and the pressure needed to make it neither too shallow nor too deep are to be learned from watching your master. If done properly, the patient will make no outcry during this cutting. You will be rewarded at this time by a river of putrefaction which slides from the opening you have made. At first it erupts under its own force; soon it slows down to a steady trickle. Gentle pressure will produce a further outrush mixed with some blood.

13. Dry oak leaves are now rained upon the wound until so many have stuck to it that they have formed an absorbent protective dressing that will be impervious to flies.

14. The stone knife must not ever be used again. Take it back into the forest, to the pit from which it was excavated. There, together with the two unused fragments, replace it gently and cover it with earth.

15. At the first appearance of the moon in the heavens, you must visit your patient in his hut. You will find him smiling and drinking wine from a gourd. Give thanks that you have been the patient's instrument, which he has used in order to regain his health.

A man is cleaned somehow in the act of making his tools. Rinsed of triviality. Exalted. The process itself takes hold. This knife that you have made—it is the only one. There are no others of its kind. It is destined for *your* hand. Feel how it rides your palm like a living haunch. Its future is foreordained. It need not compete. This knife will cut your patient well.

Witness

There are human beings who
spend their infancy in pain. One, perhaps, is born lacking, and must
be completed, or a pot of hot coffee is spilled and another must be
grafted and grafted again, first with the skin of strangers and then,
if any, his own. Still another becomes the toothsome morsel of a
tumor. This baby must be cut and suffer and even die. What is one
to think of that? When an old person dies, it is of his own achieve-
ments. But here was a body to which nothing had happened until
this pain. If this body had lived, you say, it would have known no
pleasure, remembered no comfort. It would have had only a heritage of pain upon which to base its life. It is for the best, you say!
Then what of this?

The boy in the bed is the length of a six-year-old. But something
about him is much younger than that. It is the floppiness, I think.

His head lolls as though it were floating in syrup. Now and then he unfurls his legs like a squid. He has pale yellow hair and pale blue eyes. His eyes will have none of me, but gaze as though into a mirror. He is blind. The right cheek and temple are deeply discolored. At first I think it is a birthmark. Then I see that it is a bruise.

"Does he walk?" I ask.

"No."

"Crawl?"

"He rolls. His left side is weak," the mother tells me.

"But he has a strong right arm," says the father. I touch the dark bruise that covers the right side of the child's face where he has again and again punched himself.

"Yes, I see that he is strong."

"That's how he tells us that he wants something or that something hurts. That, and grinding his teeth."

"He doesn't talk, then?"

"He has never been heard to utter a word," says the mother, as though repeating a statement from a written case history.

I unpin the diaper and lay it open. A red lump boils at the child's groin. The lump is the size of a walnut. The tissues around it slope off into pinkness. Under the pressure of my fingers, the redness blanches. I let up and the redness returns. I press again. Abruptly the right arm of the child flails upward and his fist bumps against his bruised cheek.

"You're hurting him," says the father.

The eyes of the child are terrible in their sapphiric emptiness. Is there not one tiny seed of vision in them? I know that there is not. The optic nerves have failed to develop, the pediatrician has told me. Such blindness goes all the way back to the brain.

"It is an incarcerated hernia," I tell him. "An emergency operation will be necessary to examine the intestine that is trapped in the sac. If the bowel is not already gangrenous, it will be replaced inside the peritoneal cavity, and then we will fix the hernia. If the circulation of the bowel has already been compromised, we will remove that section and stitch the ends together."

"Will there be . . . ?"

"No," I say, "there will be no need for a colostomy." All this they understand at once. The young woman nods.

"My sister's boy had the same thing," she says.

I telephone the Operating Room to schedule the surgery, then sit at the desk to write the preoperative orders on the chart. An orderly arrives with a stretcher for the boy. The father fends off my assistance and lifts the child onto the stretcher himself.

"Is there any danger?"

"There is always danger. But we will do everything to prevent trouble." The stretcher is already moving down the corridor. The father hurries to accompany it.

"Wait here," I say to him at the elevator. "I will come as soon as we are done." The man looks long and deep at the child, gulping him down in a single radiant gaze.

"Take good care of my son," he says. I see that he loves the boy as one can only love his greatest extravagance, the thing that will impoverish him totally, will give him cold and hunger and pain in return for his love. As the door to the elevator closes I see the father standing in the darkening corridor, his arms still making a cradle in which the smoke of twilight is gathering. I wheel the stretcher into the Operating Room from which the father has been banished. I think of how he must dwell for now in a dark hallway across which, from darker doorways, the blinding cries of sick children streak and crackle. What is his food, that man out there? Upon what shall he live but the remembered smiles of this boy?

On the operating table the child flutters and tilts like a moth burnt by the beams of the great overhead lamp. I move the lamp away from him until he is not so precisely caught. In this room where everything is green, the child is green as ice. Translucent, a fish seen through murk, and dappled. I hold him upon the table while the anesthetist inserts a needle into a vein on the back of the child's left hand, the one that is weak. Bending above, I can feel the boy's breath upon my neck. It is clean and hay-scented as the breath of a calf. If I knew how, I would lick the silence from his lips. What malice made this? Surely not God! Perhaps he is a changeling—an imperfect child put in place of another, a normal one who had been stolen by the fairies. Yes, I think. It is the malice of the fairies.

Now the boy holds his head perfectly still, cocked to one side. He seems to be listening. I know that he is . . . listening for the sound of his father's voice. I speak to the boy, murmur to him. But I know it is not the same. Take good care of my son, the father had said. Why must he brandish his love at me? I am enough beset. But I know that he must. I think of the immensity of love and I see for a moment what the father must see—the soul that lay in the body of the child like a chest of jewels in a sunken ship. Through the fathoms it glows. I cup the child's feet in one hand. How cold they are! I should like to lend him my cat to drape over them. I am happiest in winter with my cat for a foot pillow. No human has ever been so kind, so voluptuous, as my cat. Now the child is asleep. Under anesthesia he looks completely normal. So! It is only wakefulness that diminishes him.

The skin has been painted with antiseptic and draped. I make the incision across the apex of the protuberance. Almost at once I know that this is no incarcerated hernia but a testicle that had failed to descend into the scrotum. Its energy for the long descent had given out. Harmlessly it hung in midcanal until now, when it twisted on its little cord and cut off its own blood supply. The testicle is no longer viable. The black color of it tells me so. I cut into the substance of the testicle to see if it will bleed. It does not. It will have to be removed.

"You'll have to take it out, won't you?" It is the anesthesiologist speaking. "It won't do him any good now. Anyway, why does he need it?"

"Yes, yes, I know . . . Wait." And I stand at the table filled with loathing for my task. Precisely because he has so little left, because it is of no use to him . . . I know. A moment later I tie the spermatic cord with a silk suture, and I cut off the testicle. Lying upon a white gauze square, it no longer appears mad, threatening, but an irrefutable witness to these events, a testament. I close the wound.

I am back in the solarium of the Pediatric Ward. It is empty save for the young couple and myself.

"He is fine," I tell them. "He will be in the Recovery Room for an hour and then they will bring him back here. He is waking up now."

"What did you do to my son?" The father's eyes have the glare of black olives.

"It wasn't a hernia," I explain. "I was wrong. It was an undescended testicle that had become twisted on its cord. I had to remove it." The mother nods minutely. Her eyes are the same blue gem from which the boy's have been struck. There is something pure about the woman out of whose womb this child had blundered to knock over their lives. As though the mothering of such a child had returned her to a state of virginity. The father slumps in his chair, his body doubled as though it were he who had been cut in the groin. There in the solarium he seems to be aging visibly, the arteries in his body silting up. Yellow sacs of flesh appear beneath his eyes. His eyes themselves are peopled with red ants. I imagine his own slack scrotum. And the hump on his back—flapping, dithering, drooling, reaching up to hit itself on the cheek, and listening, always listening, for the huffing of the man's breath.

Just then the room is plunged into darkness.

"Don't worry," I say. "A power failure. There is an accessory generator. The lights will go on in a moment." We are silent, as though the darkness has robbed us of speech as well. I cannot see the father, but like the blind child in the Recovery Room, I listen for the sound of his voice.

The lights go on. Abruptly, the father rises from his chair.

"Then he is all right?"

"Yes," I nod. Relief snaps open upon his face. He reaches for his wife's hand. They stand there together, smiling. And all at once I know that this man's love for his child is a passion. It is a rapids roiling within him. It has nothing to do with pleasure, this kind of love. It is a deep, black joy.

The Slug

I am a shady man. I prefer back entrances and crooked alleys. It's what comes of an ancestry rooted in the servant class. In the matter of dwellings, I lean to caves rather than glass-walled atria. Libraries get the nod over swimming pools. Give me no wide-open spaces where the deer and the antelope play. I want to be wrapped up snug by my environs. And as for gardens, I prefer them cool and dark and abandoned. None of your primped and gussied flower beds sizzling in heat that would craze a hornet. The gardens of my heart are not found in the country, nor even in small towns and villages. It is in great cities that the oases of the spirit lie, in cities, bridged and tunneled, and with pavement plated. I do not speak here of vacant lots, which are another beloved feature of cities, but of bona fide gardens.

You do not set out to locate such a garden. It will not be there. You simply happen upon it, having just emerged from an alley where you have darted to urinate. Or, having passed a deconsecrated church now being used for a theater, you glance back and

. . . there it is! One of a pair of lichenified stone lions guards a tangle of mountain laurel, the other lion having long since kicked over its pedestal and padded off, leaving only a paw print behind.

Such a thing happened to me some years ago while taking my customary turn through the center of New Haven. Down Prospect Street I went, turned right on Grove, then left on High and . . . gleaming like Burmese jade behind eight-foot-tall wrought-iron pickets . . . a garden. Was it there always? I have passed this way many times. How is it that I never noticed it before? Ah, perhaps only now had I been made ready to see. A walk along the length of the fence failed to reveal the presence of a gate. There seemed no way to enter this place. On its farther border loomed the ivied marble wall of a great temple that was windowless. No residence this, I knew, but a Secret Society whose members alone had access to the garden. The great iron spikes were cold to the touch. Within, all was dark and silent. To me it was a hushed beckoning.

No longer monkey-agile, but retaining some stubbornness of hoof, I reached for the upper crossbar, a foot or so below the points of the spikes. Chinning, I surpassed it, and with my arms rigid and vibrating, hoisted one leg to rest between two points. I moved my hands to grip the spikes and raised the other foot. For a long moment I balanced between the skewers, risking evisceration or worse, then pushed off. At last, I was down . . . and *in.* Two steps farther and I was quite invisible to anyone walking by. How cool and dark it was. From somewhere there was the sound of water dripping on stone; each minute, three drops. The trees were thick and old: one copper beech and a grouping of hemlocks. The only light that reached the garden floor came from diffused rays that scattered through the branches to speckle the ground, every bit of which was coated with myrtle and ivy. The month was May, a fact reluctantly conceded by the blooming of a single blue iris and a scanty outbreak of lilies of the valley. It was the genteelest garden of my life. None of the huzzah of tulips, and nothing, by God, pink. Only those few pale blossoms that are the last elderly descendants of great family lines, all the ghosts of their ancestors blooming round them in the air.

I stepped between two hemlocks into a small central clearing fringed by old bridal wreath. There was a stone bench at the edge of an oval pool no more than six feet at the longest. Rising from the center of the pool, a marble nymph expectorated one drop of pearly saliva every twenty seconds. I sat on the bench and leaned back. It was still. Nothing moved. For this garden knew not hummingbird nor bumblebee nor beast of any sort. The minutes passed. I drowsed, or daydreamt, and in that altered state, thought to see the little nymph move and settle her drapery. Just so did Galatea first bend her head to Pygmalion, and smile. There was a scent in the air. Not of flowers, but something else—mushrooms or rotting leaves. It gave to this place the funerary air of a little cemetery whose single prize, having been dug up and taken away, keeps on with its vigil.

All at once, a movement at the edge of the pool caught my eye. Something had stirred. I looked and saw nothing. Again, a movement, and again I looked and saw . . . no . . . not nothing, but a *thing*—ten centimeters in length and aimed at my foot. It was a slug. I leaped up, thirty-two degrees Fahrenheit at the bone. Were there others? Was this place infested? I cast about for a stone. There was no stone. I could not have squashed it anyway. For even as I watched, the slug slid a centimeter away from where I stood. Something about its mode of locomotion arrested me.

I knelt to examine it more closely. The slug was brown as Morocco. The forepart of the back was golden with tiny leopard spots which coalesced into stripes that ran the length of the tail. These stripes seemed somehow to participate in the drag of its long abdomen. Thick as a finger it was, and full of wet stuff. Again the slug moved, and I saw the rippling lips of its single pink foot, beauteous enough to be worth all nine Muses. The head of the slug was broader than the rest, and was topped by twin antennae, each of which bore a knob. These it waved here and there, testing the air. And this slug was nude, having neither plate nor shell. I understood at once. Such beauty disdains any covering save the perfect stocking of its own integument. The enchantment it evokes is quite enough protection.

By now I was helpless in the grasp of this creature whose lineage traces unbroken from the Early Cambrian period, while I know no name more distant than my grandfather's. I was in the company of a virtuoso who had made every corner and climate of the earth its dwelling place: the sea and its bottom, pond and pool, the river and its estuary, the land, too, from treetop to cavern. And this garden. Such a slug could recall being washed to and fro by the tide, clinging through frenzies of spray to a wave-swept rock. Once, it had fanned its ancient gills and pried open oysters in estuarine glee. And this slug had listened to Plato and banished all desire. For it bears both sperm and egg within itself.

Sexual self-sufficiency induces serenity. The smug slug is complete unto himself; he is a winsome onesome. Thus, his unruffled pace. Unencumbered by the need for copulation or its human idealization, love, the slug has time to burn. Think of the weeks, months and years spent in the pursuit of love and you will guess the vast prairies of leisure that would be yours to do with as you wish were you but relieved of the need for a mate. And so, the slug ripples and arches and waves his feelers and advances an inch an hour. What does the slug care? He has only herself to charm.

Now I am beyond all restraining. That which only minutes before had sickened and repelled me was what I longed most to hold. Come what may, I must go beyond mere seeing to the closer union of touch. I reached out one tentative finger and let it rest lightly on his back. He was cool and slimy and turgid. I felt his body stiffen, take guard. I grasped him in my fingers and raised him to my waiting palm. At that first tangency, all the trajectories of my life intersected. He lay still while I stroked him from head to tail. Soon he raised his head and tried the air with his antennae. I felt the tickle of his beauteous foot, and upon my hand I saw the trail of his silvery slime.

So! It was love that I sought and found in this breathless, blind and dripping garden. But a fear swept over me, the fear of discovery. What if my children should one day find me petting a slug? Or worse—my wife? The family does not exist that would enfold to its bosom a slug lover. And what of my neighbors, my colleagues who will hear of my passion and turn aside with darkly beetling faces? They will harden their hearts to me. Soon men will come to seize me—bailiffs and sheriffs with mastiffs. Taunted and bespattered, I shall be hauled to the New Haven Green for a public squashing.

It was not meant to be. I was too old for mad elopement. I replaced my slug upon the ground and turned away. Once again I climbed those cruel bars, defied those iron prongs and vaulted from that garden as from Paradise. I have never found it again. But not a day goes by that I do not yearn for that place. For I have heard the siren song of the slug, and I shall hold that garden in my arms until I die, and thereafter.

Semiprivate, Female

Room 324 is at the end of the cor-
ridor. It has four beds, each separated from the others by a curtain
which can be drawn to achieve that status listed in the Admissions
Office as Semiprivate, Female. In this room, at this time, it is less
the presence of the curtains between the beds than the shrouded
minds of the women that secure for each her territory. In any case,
they lie hidden from each other. Only their dreams cannot be kept
separate. These issue forth in sallies from the beds to mingle aloft.
Opposite the door is a bank of windows covered by slatted blinds.
The roses are cruel that line the ledge in front. From the outside
ledge comes the low, watery trill of roosting pigeons. The nurse in
charge of this ward will strive by her presence to bind the patients
together. She cannot bear their isolation. All day she will use her
strong body, stepping from one bed to the other, spinning threads of
cheerful talk. Her hands will touch them each in turn, her voice

urge them into a community, coax. But in the end, neither she nor they will be persuaded.

All night the women have fed the sick moisture with their lungs. The curtains belly and collapse with each breath. Palpable wisps are a gauntlet through which the nurse must pass. She sets her bowed head against the air and barges in. Then goes quickly to raise the blinds. The windows are flung open, discharging the pigeons and bringing into focus the litter of the night. This nurse depends upon the magic of air and sunlight. For a long moment she stands in the center of the room and sighs for her task. Somehow she must scour it clean, make of the dark, infectious squalor a neat thatched cottage. But even then she knows that in spite of everything it would remain a stumpy, cockeyed hovel. She could hope only for the simple dignity of shiny floors and plumped pillows.

The hush of the night is still present, made even more hushed by the noises that come from the beds, as though small deer are stamping among dry leaves. It cannot be done, the nurse thinks. Although God knows she has hummed hymns often enough as she mopped and swabbed and polished. Precisely because she has tended and pitied, the desolation is hers as well. Soon she is joined by another woman, an aide. They smile and draw resolve from each other's bosoms and arms. The aide steps to one bedside and gently shakes a vanished shoulder.

"Wake up, dear," she calls into the heap of cinders and other hard particulars. "Time to get up. How do you feel today?"

A purse string is tugged, and a dusty mouth that had hung open all night is slowly gathered in.

"How do you feel this morning?" she calls out again. The woman in the bed stirs. A pair of gulls is pulling at the carcass of sleep. At last she speaks.

"How do I feel? Dead. And it ain't half bad. So go away and let me be."

"We'll have a nice bath. Don't you want a bath?" says the aide. And the two nurses begin with pans of water. Together they soap armpits and legs, taking some pleasure themselves in the warm lather, then wiping dry with towels. No crease so hidden, no bag of skin so empty, that they will not bathe and powder and cream. They turn the woman from side to side to extricate wet yellow sheets and lay new resentful ones, pulling them tight. Old dry hair that at any given cough might be dislodged from her head is brushed and braided. Now and then a groan reproaches good intentions. All the while the hungry bed curtain reaches for the backs of the nurses, mad for attention. The nurses move quickly now, emptying pans, bringing fresh water, crooning. Their uniforms stir up the smells of alcohol, tincture of benzoin, oil of cloves—sincerer smells than the cajolery of the flowers on the ledge. At last, creamed and oiled with rich vowels, the woman settles back in the bed, sighs and takes root. Who can withstand the battering? Just so, these nurses repeat over and over the simplest tasks, faithfully trimming toenails, and wiping the matter from eyes, never voicing their love, but storing it up in the bodies of these others until, one by one, they should burst into flower.

"Give us a chance, ladies," they call out to the other drawn curtains. "You'll get yours soon enough." Even before they have finished tidying, the first bed is soiled by the woman's whimpers.

But now the doctor has arrived. It is like a rock falling into moist earth. He is a big man, a surgeon after all, with a clumsy face rescued by pale green eyes and by a mahogany voice which any number of nurses have said is a powerful therapy in itself. Years ago, he had not hesitated to lower his voice even further at the bedside of a patient he had diagnosed as being susceptible. But no more. The green eyes have long since been weakened, made more permeable by the sights they have seen. He has lost an entire dimension. He has forgotten himself. Now he listens to presences. On Rounds the imperious placement of his feet has softened into a kind of lumbering. In his youth he had played football. Still, at the first sound of his footsteps the room trembles, then recovers itself.

Again and again the doctor has tried to cast his net about the secrets of this room. But he cannot. All he can do is gaze and listen. There are times when he is in a tomb ministering to phantoms. Perhaps if he were to make a diagram of this room, everything drawn to scale, if he were to count and measure every object, perhaps he would come upon the one piece that did not fit, a blunder in the design. He would study this mistake until it revealed the mystery of this place. Then it would be given to him to know. He envies the nurses to whom the patients will offer fragments of their lives, small illuminations from which by his gender or his position, whatever, he is excluded. He must divine. It is never so reliable as revelation. Once, while standing at the bedside of someone who had just slipped into coma, he had had an urge to bend close to that face, which was like a blank sheet of paper, and tell his own secrets, which he knew would be safe there and nowhere else. But he had not uttered a word to that muteness where sleep and waking take no turn, where nothing is born and nothing dies. The bed is still empty where she had lain for months blind and bandaged like an ear of unshucked corn. Coma—who can elect it? Who renounce? Her age could not have been told, existing as she had on some vestigial, osmotic plane which required only an occasional breath, only the slowest crawl of the blood. One night, even these sluggish tides had stopped. The change was imperceptible, the way rain slowly sifts through the trees long after a storm. So that when the nurse had discovered her, lapsed at the end of a long sigh, for a few moments she had not been sure.

The patient in the next bed is an emaciated woman, a Filipina in the sixth month of a pregnancy. There is a frost of ashes about her mouth. Each night she is swept clean by a fever that has burnt up every bit that is not essential—blood, saliva, tears, tissue. Only the mighty fetus, raving to be born, is not touched. Even as the child buds and splits and specializes, the woman grows daily less differentiated until she is something rudimentary, a finger of flesh, unfulfilled, unformed, that will surely die of its one achievement. She resembles a snake that has swallowed a rabbit and is exhausted by her digestion. Through the translucent, dark-veined belly, the legs of her meal, moving.

She has had no visitors, the nurse informs. One morning the sun was sparking the blinds. A breeze from the open window blew aside the smoky curtain around the Filipino woman's bed, and he had a glimpse of her lying with her robe half apart. And the flickering, radiant abdomen swollen like dough under a damp cloth. Waves of heat rose from it, distorting his vision. He was shocked to find her so young, a child, really, with eyebrows thin and arched like a moth's. And tiny inquisitive fingers probing her pouting navel. On that morning, behind the untrustworthy curtain, alone save for the ghost of pleasure she was bearing, she had been smiling! The last of it was still upon her inconsolable mouth. She had seemed then beyond the reach of anyone. He had an impulse to step through the curtains and examine her body for the imprint of a hand, teeth marks, some proof that once a man had bent there to kindle a fire and warm himself.

The doctor goes to the bed next to the one where the nurses are working. When the occupant sees him standing there, her thin neck strives to rise from the pillow, but cannot lift the heavy fruit it carries. He has often been able to tell by a woman's neck that she will not last long. The woman watches him the way a dog eats his master's face while waiting for table scraps. A flame of hope blazes up. It leaps out of her blue eyes and singes him. He finds a reason to step back from the bed. During the night a red rubber tube had crept into her nose and wound through the swollen coils of her intestine. Ever since, it has fed with long, wet sucks, until now the bottle on the floor is half filled with suds and stagnant slime. The doctor draws down the sheet to percuss and auscultate her chest. Through the stethoscope, a feeble crackling. Beneath the skin, the ribs swirl and wallow with each breath. He remembers how two weeks ago she had walked into the hospital dressed in green, which he had taken as an ominous sign. When he was a young boy he had seen the ripped lawn of artificial grass that had been thrown over a new grave to hide the rawness, but which could not conceal his embalmed father beneath it. Oh, yes, green was the color of foreboding. Life had taught him to look for these little clues. She was a timid, powdery woman who seemed to perch alongside her body lest she be overly identified with something that might not really be hers. Then, she had answered his questions vaguely, disremembering events, losing track, evading, because what she feared more than anything else was becoming trapped in this capsizing body and never being able to get out. She much preferred to hover like a ghost. When he pressed her abdomen, he knew it was her thoughts more than the pain that made her wince. Still, he saw, timidity had not robbed her of grace. Now and then, when she reached out to move something two inches to the left, her long pale fingers stirred him.

The doctor removes his white coat and rolls up his sleeves to change the dressing over the colostomy he has made in her abdomen. Her hands, trying to keep out of the way, become hopelessly lost. Beyond all intent her eyes rove the mound of her entrails. She had not meant to look. She who has always rejected all mysteries is forced to unravel this one. Not so long ago in the orange groves of Florida she had loved a man with all her heart. From the day he had left her, without a single word, she never dared to utter his name. Now, when she sees the intestine arching thick and raw from her body, she is somehow reminded. In the risen coil of bowel nailed to the surface and broiled by the unaccustomed sunlight, she sees her lover in a scum of glazed fat. The scent of oranges drifts into the room and creeps into the openings of her body.

"It is time to open the colostomy," the doctor says. "There will be no pain. There are no nerve fibers in the bowel." And he plugs in the electric cautery. It is a gun that he holds. He presses the trigger and she watches the wire loop brighten. When it is red-hot, he presses it to the bowel. There is a hiss, crackling. Smoke rises. She smells herself cooking. Then the long brown smell of feces. The doctor has long since trained himself to remain aloof from the smell of burning flesh, and presses on with his task. Still, it happens that while operating late in the evening and having missed his dinner he, too, smells the meat cooking. Then, it can't be helped, saliva gathers.

"In a few weeks it will shrink down to a rosebud," the doctor tells her. "You will only have to wear a little bag." In the harsh light the belly gasps. A fit of retching takes it. Up comes blood.

"What . . . ?" she asks him. And waits. She has the petaled look of something in a vase.

"Cancer," he says. The word blisters his lips. And he thinks of an ox trampling a pasture, transforming the grass into its own thudding excrement. And he wonders with what she would be able to persuade herself now that he has set her mind adrift in this room. He does not approve of lying to patients to encourage them falsely. But, it is possible, as every gardener knows, to fool flowers into blooming by keeping a light on all night. In the leaden bafflement that follows, he is aware of his naked forearms, the black hair on his wrists, the ropy veins. There had been trouble drawing her blood for the laboratory. She had been stuck so many times. The veins were broken, thrombosed. He himself had tried without success.

"Open and close your fist," he had told her. "It's time to feed the vampire. That's it. Pump up a good one." Her fingers bunched, but it was very far from a fist. Nor were her veins to be seen or felt—shy little lizards' tails that sensed the needle drawing near, and laid themselves flat among the hummocks of fat, changing color even. He wished he could have lent her one of his own. It was as close as he had ever dared come to taking upon himself the pain of someone else. He himself has yet to be sick. By what seems to him a miracle, considering the possibilities, his parts have continued to mesh and revolve smoothly for fifty years despite the beating he has given himself in this work. Only now and then, lately, he finds himself listening for the faint sound of turbulence in his blood, the grating of cartilage over a bony spur. That the patients did not enjoy the same physical ease, well, once in a while it did make him self-conscious, but never to the point of guilt. No one, he reminded himself, could refuse the gifts of health or disease. Still, he would gladly have opened one of his veins for her.

Now, he sees, without looking about, the crack in one of the windows, the large water stain on the ceiling, the peeling radiator. Already new blood was darkening the fresh dressing he had just done. All these things expressing themselves. Even the mop and pail, the rampant mantling curtains. He longs for his carpentry bench. The smell of the wood soothes him as he cut it to size, planes it smooth. This room is like the belly pocket of his overalls into which once he had reached for a nail and pulled out a still-fluttering brown moth.

At last the woman's lips part and the tip of her tongue surfaces. Her mouth is crammed with words that she does not know how to speak. Only the sunlight passing through the crack in the drawn curtains gives any reason to hope.

"What is the cause of it? Why did it happen?"

"It is one of those things," he tells her. And thinks of how, two days before, he had payed out the length of her intestines with his hands until he had come to the blockage, had seen, then, the peritoneum cauliflowered with growths that mounted each other wildly. Ever since her admission to the hospital she had been dosing herself with bits of the past, feeling that she had to remember it all, must not leave anything unrecollected. As though her life were a silver path that she had secreted, and in order to survive she had to retrace every step of it. Now, all at once, she knows that she will go no further than this, that she will stop here at this hole in her abdomen.

Mistakes, like copper bands, tighten around the doctor's temples. What kind of honest work leaves scabs on someone else's knuckles at the end of the day? How he hated this room—the dubious promise of the place, the duplicity of it. Toying with the hopeful, singling out the weaknesses of the downcast, until their hearts were broken. When, finally, the woman opens to scream, it is he who claps a hand over her mouth to hold it back. What comes from her is the sound of a horse blowing out its nostrils. The ceiling, having heard it all before, has grown deaf, and is intent only

upon extending its watermark that has no shape or image. At last
he leaves her alone with her hands that lay at either side of her head
to form a fragile basket for her eyeballs.

The solarium is at the opposite end of the corridor. It is another sorrow of a room. Here, the doctor watches as a man wearing
a brow of thorns waters the artificial plants with real tears. He is
the woman's husband. His name is Tom.

From then on, the woman never stops. To and fro her flesh
swims in quarter circles, paddling for all she is worth to reach a
shore. But sinking deeper and deeper. Now and then she will rise to
the surface, take a few swallows of air, squawk, then down again
to whatever murk. On her last day, at that red hour of evening
when Grace, if it will come at all, is most likely to make an appearance, the doctor, laying himself open, returns to her bedside. The
thorned man is there, holding the curtains away with his broad back
to keep them from reaching for whatever is left still quivering upon
the bed. A pot of yellow tulips sits on the nightstand and voices
assurance that nothing bad is taking place. All at once, the woman
pauses in her work, turns her head as though to listen. A breach of
clarity opens, and with it, immense pain. Her eyes are crystals of
it.

"Eddie! Eddie!" she cries out. "For God's sake, Eddie!"

But the husband's name is Tom!

"Shh . . . shh," he says. "I'm here."

"Eddie! Oh, love!" she calls out again. She is running wildly
through streets of confusion, thinking to tag a footstep, making a
last run for it.

"Yes," says the man simply. "It's all right. I've come. I'm
here." And takes her hand. But the hand will not be deceived and
flicks away an imaginary insect. So that the man, Tom, knows that
she has kept a secret. He wrenches his gaze to a distant place
somewhere above the bed, looking after something slippery that he
had once held but that had suddenly spurted from his hands.

A young boy, too, leans against the gassy curtain. Dazzled, the
boy cannot pry his gaze from his mother's face. Nor ever would be
rid of it. Years later, in these deep breaths, one after the other,

rolling in, rolling out like the sea, in the deepest of them, he would locate his own most painful memories.

Much later, the doctor hurries up the steps to his house. He turns the key in the lock, opening the door only wide enough to let himself in. But it is no use. Like an unwanted homeless dog the room follows him into the house, demanding that he step into it and *do something*.

On Hospitals

Last night, a robust night. Tequila, pitchers of beer, with a pack of young poets at a saloon called Viva Zapata. Much laughter and recitation; even the bartender there is a poet. I would have much preferred wine but, if one is romping with poets, one can scarcely expect to be wheedled out of his faculties like a gentleman. At midnight I looked about to discover that I was the oldest living human being in the world. A twinge of shame. I should be at home, I thought, in my slippers and my bronchitic scarf. I am no Socrates that I should play jacks in the streets with the youth of the city. And a doctor, to boot!

At the hospital this morning, to find that Sylvia Taylor has died during the night. Her seventh postoperative day. Massive pulmonary embolism. What a spiteful thing is a hospital. Insofar as buildings have temperament, a hospital is passionate and jealous. All doused with prayer and gilded with blood. On Rounds, I step into Sylvia Taylor's room, now empty, the bed so very *made*, its sheets pulled rigid, the corners fiercely tucked in, the nightstand bare. This room where such immense anguish has played itself out, not just this once, but over and over for a hundred years. The impress of supplication is upon the walls; the scent of Sylvia's pain lingers in the draperies. The room is like a ruined temple shorn of all its regalia, mute now and haunted, and keeping within it the simple incandescent truth: Sylvia died while I was out carousing. Oh, the anguish of guilt! Gethsemane had not been emptier than this room with Sylvia gone from it. But do not, in this one instance, follow in my footsteps. Excessive guilt is a kind of suicide. If the Oil of Mercy is not to be found, if the drop of it you seek is not granted, why then give up your dead to their demolition. Do not clutch them endlessly to you, nor be like the pelican who, having searched and found no food for his young, rips open his own breast to feed them on his entrails. Accept the fact of death. Read the biographies of all the great doctors—Osler, Pasteur, Cushing—and on the last page you will read, "And then he died." Methuselah lived for nine hundred and sixty-nine years; he begat sons and daughters. And then what? And then, he died.

> Huddled in dirt the blust'ring engine lies,
> That was so great, and thought himself so wise.

Still, I know, I know, there are the nimble dead, so etched in my mind I could resurrect them from their dust. Men and women so poor than even their thoughts are threadbare, children in dark rooms reaching out their arms to stir the darkness.

Once I was present as the wrecking ball slung into a majestic old hospital where I had worked. A new one, all of chrome and steel and prestressed concrete, was to take its place. But I could not watch. At the first crash of the ball I fled from the scene. Who can behold without horror the eviction of so many ghosts? Why then, you ask, do I not leave this hospital that has caused me so much pain? Because, even if I should cut myself off from it, I would feel the unabated pain of the amputee, who hears the footsteps of the phantom leg in his brain, and suffers on.

Besides, the hospital is to me as Notre Dame to Quasimodo

the Hunchback. It is my egg, nest, house, country, my universe. I know its underground tunnels, its crawl spaces. Fire escape and dumbwaiter know me. I straddle the great effluent pipes that move along the ceiling of the morgue and feel the human race flow out between my legs to the sea. I crouch in the refrigerator among lumpy bags of blood ticketed for the next day's infusions, listening to them gossip about their donors, conjecture over their recipients. Here is a building that I resemble, that looks like me. We are blood relatives. I love it—belfry, parapets, gargoyles and all. No slug who makes his shell from his own saliva, constructing his house from within himself, is more at home than I in my hospital.

Doctors and patients do not alone a hospital make. Look about you at the others who labor in the same vineyard.

I had a friend who looked like Kafka—the same triangle of a face in which every bone was visible. The same dark eyes, sharp nose and pointed elf's ears. Gerald had a hollow, strained voice, as though he were shouting into a wind. I think that his vocal cords failed to approximate completely so that his glottic chink was always a bit open and there was nothing to brace the air against. His r's, w's and l's were in some disarray.

"I can't wead," he said. "I wike to wook at pickchews."

For twelve years Gerald scraped trays in the hospital cafeteria. Dressed in soiled white pants and a white shirt, and with a great paper cylinder resting on his ears, he stood behind the tiered shelf where the trays were stowed. He would pull the trays in toward himself, remove the dirty dishes and scrape the garbage into a trash can. I don't know how it was that Gerald and I began to have lunch together every Tuesday. As Emily Dickinson said, "It was just a happen."

Gerald knew and remembered everything about me—what I had eaten for lunch the previous Tuesday, where I was supposed to be at any given time, what operations I performed—everything. He had his sources. Such details are unimportant to you and me, but to a lover they are crucial.

"What did I eat last week?" I would ask him. "I don't want to get the same thing."

"Clam chowder, Brussels sprouts and cherry pie," he would recite, and his face would brim with the joy of such knowledge. By just such intimacies had Gerald gained his power over me.

"I'm what you call mildly retarded." He announced this importantly, the way, I knew, it had been explained to him.

"You look tired," he would say. He was given to non sequiturs.

"I *am* tired," I would agree. He would roar with laughter at this.

"How old are you?" Gerald asked me. "I'm thirty-one." He looked thirteen. A small thirteen.

"Come on," he coaxed. "I told you. Now it's your turn."

"Fifty-one."

"Fifty-one!" It was a number so hilarious as to double Gerald up and send him sliding from his chair.

"What's so funny about fifty-one?"

"You don't look fifty-one. I weigh seventy-eight pounds." I saw that he did. Gerald was six inches shy of five feet, and eel-thin. His skin was devoid of color, white and bloodless. His hair was straight and black and unevenly planted, the way a tear in upholstery will spill out some of the kapok here and there. Each of his ears was missing a ridge of cartilage, which gave them a thin, unrolled look. The nape of his neck remains one of the most touching anatomical events of my life.

When Gerald wasn't worrying about his health, he was smiling. And he could be easily distracted from his hypochondria. He adored presents and accepted anything, no matter how meager, with a joy that passeth all understanding. Nor did he once say thank you, which is much different from being thankless. It is merely a different concept. To Gerald a thing given was a thing found, a stroke of good luck. The giver was the blind instrument of Fortune. I gave him everything I carried that was not essential, in order to collect for myself another dollop of Gerald's joy. It was wicked of me, I know. But would you condemn a man for taking what he really needs?

Once in a blue moon, I took Gerald to a saloon near the hospital for a couple of beers and a few cigarettes. He would make me promise to bring a medical textbook so that we could look at the pictures together.

"What's that?" he would ask over and over, pointing to a picture. When I told him, he would gargle in disgust, or laugh. Once I caught him studying with ill-concealed hunger a picture of the female pudenda. I cannot bear that kind of nakedness. I thought of taking him to a whorehouse, but there aren't any in New Haven. Anyway, I wouldn't know what to do there either.

One day Gerald came to my office for a visit. Just dropped in. I gave him a pair of my shoes that I had just had resoled.

"Sit down there," I told him, pointing to the swivel chair behind my desk. He was thrilled. I knelt and unlaced his shoes and slipped them off. I put my shoes on him. They were a vast acreage about his narrow raccoon feet.

"How do they feel?"

"Good," he said. "They feel real good."

"They're not for work," I said. "They're just for you, at home."

"No," he said. "Not for work." We looked at each other, then lowered our gaze to his feet.

"You're following in my footsteps," I told him.

"I'm wearing your shoes," he said.

"I have to get to work now," I said. Gerald smiled and clomped out, barely able to hold his new shoes on his feet.

One day I gave him my sweater.

"Put it on," I told him. He did, awkwardly, his hands going all over the place.

"I'm no good with my hands," he said.

"You look good in green," I replied.

Gerald loved to be told a story. Best received was a long account of a painful illness terminating in near death from hemorrhage or rupture of this or that. The hero was always a surgeon who operated on the patient just in time. Gerald would sit through my bloodiest improvisations, rocking back and forth, his all but nonexistent buttocks working against the chair in inexpressible relishment.

"You are a dear man," he would say when I finished. "You never change." It was the closest Gerald ever came to confessing his love. I would strike a Great Surgeon's pose and out of the corner of my eye watch him *liquefy*.

I did not see Gerald for a month because I had undertaken to teach a noon-hour class. Nor did I see him the next month because I went on vacation. When I came back, it was to find that Gerald had been admitted to the hospital with pneumonia the week before. Within three days, he was dead. I went to visit his mother.

"He missed you," she said. "He was pining."

You must never run out on an elf, or hurt an elf's feelings. To an elf, such a wound is invariably fatal.

Yet, here, if nowhere else, locate mercy. The people who work here have forgiven every odor and leakage of the flesh. They would take every risk to plump a pillow, give an injection. Rightly, they are dressed in white.

There is a scene that still, now and then, drifts among my thoughts despite the thirty years since I beheld it. It is before dawn. I have just stepped for the first time onto the wards of the Albany Hospital as a student. There are two nurses—large, pillowy women of a type not prized in this time when tendon and bone are à la mode. They stand on either side of a bed in which a man lies, crumpled. During the night's agony, one cause of which is ulceration of his legs, he has worked his way down toward the foot of the bed where now the wreckage lies knotted and helpless. The two women bend above him, take him at each armpit, their other hands meeting under his back. "One, two, three," they count aloud together and haul the man up to the head of the bed. He is very heavy. But these women lift him with their breasts and their eyes as well. Even the intake of their breath is used to hold him aloft. Caught slung between them, he is Christ being taken down from the cross.

On Hospitals

A cart stands nearby upon which a metal washtub steams over a gas burner. One of the nurses stirs the contents with a pole, then fishes out a towel heavy with water. The two women wring out the towel between them, flap it once or twice to cool it a bit, then wrap it about the blue, ulcerous leg of the man in bed. In the gray, boisterous ward, the white towel with its aura of mist floats from the hands of the nurses to settle about the man's leg and all but smiles, keeping its mystery. Instantly the wet heat does its work. The man sighs with contentment. His pain recedes.

Ever since, I have depended upon the kindness of warm compresses. I suspect that wet heat applied to human flesh has done more to ease the plight of the sick than all the surgery past and all that is to come. A compress gives such an insinuating warmth. It cooks the tissues gently, giving off a humid aroma, softening resistant fibers, drawing out poison, polishing and smoothing all the rough places of the body, much as footsteps and knees hollow the stones of an old church. Warm compresses are sensible. Anyone knows that to rekindle a near-cold fire, you place one faintly glowing ember against another, fan it for a moment, then sit back and watch a pretty flame break.

Chatterbox

———— •••• ————

Homage to Saint Catherine of Siena

It is not always the doctor who heals. Sometimes it is another patient.

The following is taken verbatim from the diary of a woman who was a patient of mine some years ago. It is in the form of a letter to her brother Raymond. The diary was mailed to me by her brother some months after her death.

Catherine Goodhouse had been what in high-school yearbooks is called a "chatterbox," and in textbooks of psychiatry is called a "compulsive talker." Catherine was a talker the way other people are writers or actors. Talking vouchsafed her life. "Do you hear me?" she seemed to be saying. "Then I am alive." It satisfied her as sucking satisfies the baby who likes the feel of his lips and tongue working.

Talking was an affliction of which life had made her painfully aware. One day, her husband Joe had stood up in the middle of Sunday dinner and walked out of the house. She never saw him again. When he was sixteen, her son Warren had said to her, "Why the hell don't you shut up?" Then he was gone—enlisted in the Navy. For four years Catherine lived alone, talking to whoever would hold still long enough, or to herself. She couldn't help it.

When I first went into practice, I used to work part-time at Golden Gardens to make ends meet. The Nursing Home was far enough outside town to enable the owners to describe it in the brochure as "rustic" and "rural." What the people in Golden Gardens needed was not so much country air as someone to talk to. Company. All day long they sat encased in silence, unable to find a way out of it.

"You ought to sign up at Golden Gardens," I told Catherine. "Go out once or twice a week to visit with the patients. They're lonesome, and you have what it takes." The next day Catherine called to say she could go every Tuesday and Friday from two to three in the afternoon.

Catherine Goodhouse was one of those petite, doll-like women given to lavender sachets and dresses made of organdy or dotted swiss. Every time she came to Golden Gardens she looked and smelled like spring itself. The patients adored her. She was assigned five of them, the same five each time, so that they would come to look upon her as family. There were four women and a man. Two of the women were in wheelchairs. When she arrived at five minutes to two, they would already have been assembled in the "parlor," and had been drawn up into a little semicircle in front of which Catherine would stand babbling on about what she was wearing, where she learned to bake, what she was going to plant in her garden. The weather. It was only fair, she told an attendant at the home, to start with what she was wearing because Mr. Freitas and Mrs. Celli were blind. As far as anyone knew, none of the five was ever heard to utter a word in reply. It didn't matter. They just sat there for the hour, rapt and adoring, as though she were the most beautiful, most charming, the freshest thing in the world. As time went on, it was plain to see that the five old people were stronger and more alert than they had been before. They certainly looked better. Their faces were pink, their lips moist. Pretty soon, Catherine was going to the home five days a week.

"Well now," she would say at the close of each of her visits, "we've had a lovely time, haven't we?" And she would walk slowly toward the door. All five, including the two blind ones, would turn to follow her right up to the last second. About a year later, I stopped working at the home, and lost touch with Catherine. Three years later I heard that she had died. And then there was this diary.

My dear brother Raymond,

For some time I have been going out to a convalescent home to talk with the patients. They don't have much company, you know. I cannot believe now that when I first saw them I was horrified. That first day and for a while afterwards, they seemed not only grotesque but dangerous. Were they contagious? I wondered. Or violent? And with all their bodily functions—salivation, excretion—unpredictable, sort of wild. If one of them touches me, I thought, I'll die. There was Mrs. Greenwald, with her wet chin and that cackle that just welled up out of her every little while. And Mr. Freitas, with those boiled swollen hands that looked like blutwurst. His fist on the handle of his cane was like a just-born infant's face, engorged, angry. His hands seemed always to be hot. He would reach for things to cool them on, like the armrest of his chair or the knob of his cane.

But right from the beginning there was something about them that drew me. It was their pathetic confidence that they were human. I saw bravery in their very act of coming to listen to me. And they never missed a visit. I knew that they needed me as much as I needed them.

One day there were only four of them. Mrs. Greenwald was missing. "Where is Mrs. Greenwald?" I asked. Of course, they didn't answer.

"A massive hemorrhage from the stomach," one of the nurses explained. "She was dead in an hour." A hemorrhage, I thought. As though the old woman's blood somehow knew it was dwelling in a doomed body, and had hurried to escape it. I swear it seemed that the other four were embarrassed that I should have to find out and be upset by something that one of them had done. All at once I felt bereft, deprived of something, like an old coat that, without knowing it, I had been wearing to keep myself warm; and now that it was gone I was chilled to the bone. From that moment I never had any doubt that I would be going to Golden Gardens for the rest of my life.

And now, Raymond, I want to tell you as well as I can what happened. On the day of the event, I bathed and dressed carefully. I had bought a new white dress, and I was wearing it for the first time. White is so virginal, the clerk in the dress shop had said, and she had smiled knowingly. And she was right. It did make me feel . . . well . . . cleansed to put it on. I don't know why I am telling you about that except that now it does seem important to me somehow. It was such a good visit. I told them so many things. You should have seen how they were nourished by my talking. I like to think that they would have been dead long ago if I didn't talk to them like that. I had been told to limit my visits to one hour so as not to tire them, but on this day about which I am writing, they didn't seem a bit tired at the end of an hour, and truthfully, I hadn't had enough either. I just went right on for another hour, while the four old people sat there and listened.

At the end of the second hour, I stood up. "Well," I said as usual, "we've had another lovely visit." First I brought the two women in wheelchairs out to their rooms. The third woman followed on her own. When I returned to our meeting place, I saw that Mr. Freitas had stood up but had made no move to leave. "So, Mr. Freitas," I said, "it's time for me to go home." Still, the old man made no move to leave. Instead he took one step closer to me so that he could almost have reached out and touched me. Another step brought him still closer. We were standing facing each other. All at once I was aware of a strange, warm sensation. As though my body were darkening, and even my dress were pink. I could have heated the building with my presence. Thus we stood for a long moment. I studied the old man. Every bit of his clothing took on an immense importance, just as my dress had seemed to me important before—his stained vest, the yellowed armpits of his shirt, the crumpled stovepipe trousers. It seemed to me special raiment. Even the angle of inclination of his head had meaning.

Then slowly the old man raised his arms and held them out. If I were a clock, he would have been pointing to ten minutes before and ten minutes after the hour of twelve, which was my throat. My gaze was lashed to the dance of the calcified artery at his wrist, which beat at the same rhythm as the pulse clapping in my ear. As though our blood had hurled itself across the distance between us, receding from me to flow into him, then back again. Even so, I had the most ordinary thoughts. I wondered, for instance, what direction I faced. Was it north? No, impossible. It was east. Yes, I was facing east. I remembered something that had happened a long time ago when I was a young girl. A boy at school had approached me. His hand had been made into a cage. Through the aperture formed by the curled index finger and the web of his thumb, I could see the head of a little bird. A purple finch, it was. The bird lay still in the firm grasp of the boy. Beneath the row of fingers which dipped into his palm, I caught sight of three tiny talons and a tiny foot.

"Would you like to hold it?" he asked me. I felt desire and revulsion at one and the same time. I *wanted* to hold that bird more than anything else in the world. And I did *not* want to hold it with exactly the same intensity.

"Here," said the boy, "take it." I held out my two hands, covering his fist with both of mine, and I received the little bird. In the quick movement of departure of the boy's hand from mine, I felt the fluttering of the bird's wings. All of a sudden, I became terrified, and I screamed.

"Take it back. Please take it back!" He did, relieving me of it as gently as possible.

Now I stood in front of the old man, feeling the same confusion of desire and dread. My blood was tumbling. I felt that I must leave at once, or die.

"I must go," I managed to say. But I made no move to leave. He was like the stub of a white candle melted, run over, but with the flame undiminished. With each intake of my breath, that flame bowed toward me, designating me, electing me. I could feel its heat upon my cheek.

Suddenly, I was afraid. Perhaps he meant to throttle

me. Do it! I thought. Do it! Whatever it is you are going to do, for God's sake, do it! Slowly his hands enfolded my neck, closing upon it with a perfect balance of lightness and firmness, as though it were a small bird. And when, at last, his fingers reached me (they were so long in coming), I slipped into them, was caught, lay still, quieted, the way a bird, once caught, ceases its frantic efforts to escape and embraces captivity, understanding at last that captivity was what it had wanted all along.

His opaque blue eyes, all scarred and milky, were fixed upon me. They took in no sight, I knew, but seemed now to give forth a light of their own, as though far behind them they kept an everlasting source of it which needed no ignition from the outside. And just as a sailor *confides* in a lighthouse, so did I confide in this light, certain that long after he left me, his eyes would glow on. His eyes contained the world. They conjured me. I existed in those eyes.

What a feeling of utter renunciation I had! I would not have held anything back from him, would have surrendered blood, breath and whatever else he would have. I remained still, listening to the singing fingers about my neck, and my heart dwelt in a snug cottage. For his touch was hushing, the way a finger to the lips is hushing. With just such infinite gentleness the old man bade me be quiet. And I was quiet, silenced, healed.

For a long moment the old man held me between his palms. At last his grasp lightened and he withdrew. The shuffling of his feet and the tapping of his cane told me that he had gone. I did not dare to turn and look. Alone, I had the feeling that something had at long last been set right, as though a room full of awkward, ungainly furniture had been rearranged by someone with style and good taste. Long after the old man had left me, I stood honoring the event, letting comfort fill me up, expand in me the way dough expands into bread or sleep swells a lover's face into softness. Had he really touched me? I wondered. Had it really been his flesh upon mine? Or merely a close proximity, something that had come careening from a great distance to swipe within a hair of me, then dash away? But I knew that he had.

I still go to Golden Gardens to visit. But it is different

now. Sometimes, I don't talk at all. Sometimes I just sit with them. I bring flowers and arrange them, or cookies that I have made. These things they receive with the same pleasure as my talking. Now and then, one of them will smile happily at me the way a mother does whose child has been ill and has recovered.

The Virgin and the Petri Dish

―――――――・••・――――――――

Conception. It is something closed to men. Wherein a woman becomes a vessel, a harbor, folding her body about a jot of yolk and protein. She is an enclosure now. Unable to conceive, a man marvels: Were all the energy of conception to be harnessed, he thinks, it would be a cataract that could fuel the world for all the time to come. A woman smiles to herself and thinks: And so it does. And so it does.

First there is the laying of the egg. Each month, at just the right time, when the hormone runs highest in the blood, an egg is extruded from the surface of the ovary and caught up in the fronded mouth of the Fallopian tube. By the soft undulation of membranes and the waving of a million tiny paddles, the Fallopian tube propels the egg the whole length of its canal until, at last, having brought the egg to the horn of the uterus, the little pearl of great price is relinquished. Like an enclosed garden in which the rarest of plants is to be grown, the womb has been raked clean of all weeds and debris, and made ready to receive the egg.

All these things are done in the silence and darkness of the woman's body, while outside, the sperm carrier waits, biding his time, dreaming of the day when he will let out the equipment on which he dotes, and in which all of his pride is invested. At last, the time arrives. A signal is given—perhaps a glance, a word, a certain touch, and he knows that the time has come. The woman knows, too, for the hunger of her womb has become voracious. With a pounding heart and immense elation, the sperm carrier positions himself just so. Even as the rest of his body grows hard, his heart is softened by love. For what he is to tender unto the woman is his own fierce urge for rebirth. All at once, the sluice is opened. There is the spurting of a hundred million sperm of which only one will claim the egg. How profligate is the sperm carrier.

By the deliquescent fusion of two bodies, each blends its separate biology and history in the impulse to carry forward into the future. It is an instant of pure physical happiness, a moment so intensely felt that it must be forgotten immediately, or remembered only as a vague and distant contentment. One cannot live with the acute awareness of ecstasy. That which is Paradise for the instant would become Hell for the hour, killing the participants in a fierce and unrelenting joy. Orgasm is not unlike the pain of childbirth, which must fade from the memory of the one who has endured it even as the newborn infant slips free into the outside world. To remember the agony would surely result in hatred instead of love for the child. And so, the woman receives the baby at her breast, and forgets her labor. She is transformed by the miracle of genesis into a mother. It is the time of her miracles.

After the ecstatic discharge in which the sublime notion first occurred, after that secret gift to her body's inmost part, the hunger of her womb is sated by a single, whip-tailed jot of protein. Once fed, the woman becomes the feeder, hung with the heavy fruit of womb and breasts, nourishing the prize with her blood and oxygen, changing her shape to accommodate her cargo, and thriving to her high purpose. All this while she is unaware of the minute and precise shaping that is taking place within her, the pressing and flowing of cells, the rising and falling of ridges and mounds, the scooping and tunneling and arborization that is the development of the embryo. Quite blindly she dotes upon the live thing whose shaping retells the whole history of animal life. From the single-celled protozoan through the hydralike colony, through fish, amphibian, reptile, to pig and primate, the embryo passes through the phases of evolution as though to remind the woman of her own beginnings, and to establish her place in the stream of life. By this process all the wisdom and love of one's ancestors is gathered and directed into each individual.

It is like the creation of a work of art. More, it is a reenactment of the Creation itself. The Creation, wherein the mountains and valleys, the rivers and seas, shifted and flowed into new shapes and positions until, satisfied at last, the Creator let the world be. And all done with a rhythm that is ancient and a vigor that is fresh and new. The furled fetus neither sleeps nor wakens, but fastens upon its mother's blood, sharing her dreams, her wakefulness. Until, at last, what the woman has created is a child, her child, with all the parts of every other child, but never to be confused with any other child. This child that can be born and die anywhere, but can be conceived only within her womb, and whose release from her body is akin only to the release of the soul from her body. It is part of the same thing, teaching that birth and death are the same process. In the woman's undertaking, her task, what comes to her assistance to let her be so divinely exact? In this, as in every art, there is an inner core of heavenly precision.

The present spate of technique manuals notwithstanding, there seemed to be but one way to perform the act of procreation. Or so it seemed until the discovery of the test tube (actually, it is a petri dish) as an alternate conceptacle to the womb. Oh, there have been some notable variations. The case of Danaë, for instance. Danaë was the daughter of Acrisius, king of Argos. About her it had been prophesied that the son she would one day bear would slay his grandfather, Acrisius, and take over the rule of the kingdom. Acrisius imprisoned Danaë in a bronze tower to keep her nulliparous. Now it came to pass that Zeus discovered the beautiful prisoner, and he lusted after her. Unable to approach her in order to consummate his affection, he did the next best thing. Zeus descended upon Danaë in the form of a shower of golden rain, during which drizzle, Danaë conceived. The son she bore was named Perseus, who, as everyone knows, went on to slay monsters and rescue beautiful maidens. Not only Perseus, but Venus, of the sea foam born, and Athena, sprung full-grown from the brow of Zeus, were the products of rather queer conceptions.

In the animal kingdom, a kind of conception can sometimes be achieved without the addition of a sperm to the egg. This is called parthenogenesis. Scare an ovulating rabbit enough, and she will simply begin developing an embryo from her egg alone without benefit of a sperm. The rabbit born of such self-containment is, however, sterile, and cannot reproduce.

Hallowed beyond all other conceptions is the Virgin Birth. One dwells upon it with wonder and reverence. The meaning of the Virgin Birth seems quite clear—that Jesus, in order to accomplish his mission, must appear among us impollute, untarred by the Original Sin with which the rest of us are brushed. Much has been written of the womb and hymen of Mary. It was said that her womb was like a well covered over with a chained lid in order that the water therein be kept pure; elsewhere, she is a sealed fountain. Long before the Event, Solomon in the Song of Songs foretold of a closed and sealed garden in which something special would grow. In the many paintings of the Annunciation, Mary is shown seated or kneeling in her room. She has been reading a book or praying. On a table nearby is a vase of lilies and water. Through the windows of the room streams a golden light. And hovering nearby, with his head leaning toward Mary's, the Archangel Gabriel. *"Inclina aurem tuam . . ."* he seems to be saying. "Give me your ear." And then, the thrilling pronouncement: "For lo, thou shalt conceive and bear a son."

One knows at once and beyond peradventure that this is the actual moment of conception. One accepts and is made reverent by the representation of Mary being impregnated by the angel's words whispered into her ear. Impregnated by a word! Some of the more explicit paintings show a tiny scroll about to enter Mary's ear, a scroll upon which is written the prophetic command. The ecstasy upon the face of the girl confirms the miracle. To the infidels and skeptics who would demand proof of the Annunciation, I would point again to the painting, to the place where the golden sunlight is streaming through the glass, yet leaves the panes unbroken. If sunlight can pass through glass, leaving it intact, then why not the Holy Spirit into the body of a girl, leaving her *virgo intacta?* That the rest of us procreate not by words, but by the energetic fusion of our bodies, that it is a bed rather than a sitting room that is the preferred conceptory, and that it is at night rather than during the daylight hours that we are most conceptious—all these distinguish us from the gods, and place us squarely among the mortal creatures.

What a far cry from the Annunciation is conception maneuvered in a dish. In the one, there is pure spirit, in the other, pure technique. In test-tube fertilization, no archangel, but a gynecologist in sterile regalia attends the woman; there is no scroll of words aloft at her ear, but a laparoscope to be thrust through the wall of her abdomen. Still, prayers accompany both occasions. In each, the hand of God is manifest.

Say that a wife is barren. Say that she is healthy in every way save that her Fallopian tubes are closed to the passage of the egg from the ovary to the uterus. Some long-healed infection, perhaps, that laid down scar tissue to block the way, or else a congenital malformation of the tube. For whatever reason, the egg will not pass to the uterine plain where the restless army of sperm is milling about. The woman and her husband have been unable to procreate. That he, the sperm carrier, may be "at fault" in their infertility is well-known. His hundred million strong may be no more than ten million. Are the few fit? one wonders. Might many be huckbacked, others crook-tailed? In the curdling of milk into cheese, the thicker the milk, the better the cheese curdled from it. So spake Aristotle.

Now the woman engages a gynecologist who measures the hormone levels in her blood, examines her temperature chart and obtains an image of her ovaries with ultrasound waves. By elevations in hormone and temperature, and by changes in the size and shape of the ovary, the gynecologist determines the precise day of her ovulation. On that day the woman is admitted to the hospital, and taken to the operating room. Here, under the same sterile conditions that prevail during any operation, a lighted tube, the laparoscope, is inserted through the wall of her abdomen. The peritoneal cavity, with all of its organs themselves sleeping like a litter of unborn babies, is now illuminated. The gynecologist aims his scope at the ovaries. The telltale rising of ovulation is observed. Through the instrument the little follicle is entered and a single ovum captured. This is called harvesting the egg. The laparoscope is withdrawn, and the small incision in the belly is closed with a few sutures. The harvest is placed in a petri dish, a round glass dish with a fitted glass cover. In this dish, the ovum is treated with the ejaculate of the woman's husband. If Heaven is looking over the woman's shoulder, a sperm will enter her egg, and fertilization will take place. Hold your breath in wonder! It is the moment of Conception.

The embryo, for such it is now, is placed in safekeeping for thirty-two hours. During this time it remains in its dish and the single cell will divide into two cells, and they into four, then eight, and finally sixteen cells. Now the creature has the appearance of a mulberry, and is called a *morula*, the Latin word for that which it resembles. At the sixteen-cell stage, the tiny product of conception is drawn up from its dish through a tube and instilled into the woman's womb through the neck of that organ. Once again, if she is favored, her *morula* will tumble to rest upon the rich bed of her womb. There it will adhere and attach itself. This is called the Implantation. The out-of-body experience of the embryo has come to an end. Its brief foretaste of worldliness concluded, the fetus dwells within its own mother until nativity.

By just such techniques are the embryos of prize cattle implanted in the uteri of rabbits and shipped round the world to be reimplanted in waiting cows. It is a topsy-turvy world that sends a rabbit to do a bull's work. But, why, you ask, is the embryo moved from dish to womb at sixteen cells? Because, you are told, that is when Implantation usually occurs "in nature." But why not leave it in its dish until it is thirty-two cells? Or sixty-four, or one hundred twenty-eight, and so forth, transferring it to larger and large petri dishes until at the end of nine months a full-grown roseate *baby* is presented to the happy parents on a petri *platter*? Why not, you ask?

"Stop asking so many silly questions," the gynecologist says. Still you press him, until at last he admits, "I don't know. No one has ever tried it." Well might you gasp. Given the diabolical curiosity of Science, you are quite certain that one day we shall all find out.

"No ethical problem here," says the gynecologist. As there may be with artificial insemination, say, or abortion. If procreation is desired, and if the ideal of parenthood is that each parent should love the other at the moment of conception, and that each parent should love the one conceived, what does it matter that for a brief period of time there is an extracorporeal antechamber of the womb?

"Besides," says the gynecologist, "it is only a matter of plumbing."

Maybe so. But questions of wonder and dread arise. From such impious rites will we come to know naught but unhallow'd joy? Will the child so conceived, so carried, lifted painlessly from his dish and handed to his parents—will he one day ponder these affairs and think of himself as but half-conceived? Misbegotten? Doctored terribly?

Will the manipulators of genes team up with the gynecologists to form a new industry, the sinister directors of which will be called Manufacturers of Life? These directors, instead of permitting us to go on cherishing our differences, will they persuade us to meld until we are, everyone, as handsome and brilliant as the other? And all, then, of a mind to enslave or massacre those left otherwise?

And what of the memoried past to which each of us is heir? If we are to believe the poets, philosophers and psychoanalysts, our embryonic lives, though unremembered, are there to be drawn upon in moments of inspiration or anguish. Will the dishborn human being, in a moment such as this, long for his half-remembered petri dish? Will he take to living in round glass houses with round glass roofs? And will the artists of some future day, moved by adoration, paint their versions of *The Harvest, The Fertilization* and *The Implantation*?

Instead of an archangel holding a lily and whispering, they will show a masked gynecologist holding a laparoscope; instead of Mary, kneeling, ecstatic, a covered glass dish; instead of sunlight streaming through a window, no window at all through which germs might permeate. For it will be sterility rather than purity that is painted there.

In the end, praise Ishtar. What does it matter, if at the last, a huddled child awakens, stirs and moves from his world-within-a-world outward, toward companioned love and the sun?

The Grand Urinal
of the Elks

———————◆•◆•◆———————

In the beginning urination was an outdoor event. Adam distended, unleafed his spigot and let splatter wherever in the Garden he happened to be. No bush nor boulder nor tree was essential as prop. Windblown, augmenting the rain or reflecting the gold of the sun, urine was voided with the freedom and openness of beasts. And with less purposiveness. Unlike the loping wolf who hoards his water, doling it out here and there to mark out his sphere of influence or to attract sexual partners, man holds his piss in no such concern. He does not use it, but would rid himself of the whole lot of it at a single standing, then get on with the business of the day. No fellow human reads his chances from the ammoniacal fumes of a rival. McEnroe does not pause to sniff what Borg or Connors has deposited on the other side of the net. It is in the nature of man to spend his urine with the generosity of the guileless. And indeed, urination is the most innocent of acts. Dangerous, too. For, exposed, standing still and engaged in no other preoccupation is he never so vulnerable from the rear. So stabbable, so brainable, so garrotable.

Let me pause to make clear that upon the subject of female urination I profess no expertise. My own first encounter with female urination was an inadvertence which took place at the age of five, when, having wandered from a family picnic, I caught sight of a girl of my age, squatting over a bed of wild violets. This convinced me that females did not urinate; they expressed little lavender bubbles. A concept that I relinquished only years later when, perforce, the evidence became incontrovertible.

Time was when urination was a wholly subjective process. One considered not at all where one did it or who might be looking on. It was only much later that man gave up his yellow liberty and consciously sought out specific objects upon or behind which to void—*a* certain tree, *a* bush, *a* rock. Thereby going from "I peed" to "I peed on a bush, rock, tree, etc." The element of objectivity was introduced. Was it an awareness that he needed some concealment, some protection, that caused man to urinate *behind* something? Surely it was not out of a sense of modesty. Before long, what had begun as a measure taken for safety became a way of life, part of the mores of the tribe and incorporated into the heritage. The proper place to urinate was passed on from one generation to the next. "Piss here," instructed father to son. "You must use the pissing rock." Rock training, if you will. Soon it became the wisdom of the ages. One expressed contempt for an enemy by urinating on his father's grave or upon his doorposts. Physicians became piss prophets, diagnosing disease by the inspection of urine.

Having a specific receptacle for his urinary stream gave man a sense of coziness that urinating into the wind could not impart. There was a man, and there was his target, only to catch sight of which was to arouse the desire to void, the object thereby exerting its subtle influence upon the subject.

Forty years ago, as a newspaperboy in Troy, New York, I had occasion each day to visit the local Elks Club. Just off the lobby was a slatted swinging door which concealed the torso only of the passers-through. On the wall above this door a board had been nailed bearing the ancient Latin misnomer *lavatory*. Although the term is not gender-specific, it was understood by one and all that this was the place where men went. One simply knew. It was part of being an Elk. The Ladies' was secluded at the end of a hallway where the sounds of feminine physiology would be safe from the ears of the village urolagniac.

Just to pass through this swinging door was to enter a different world. It was the Baths of Caracalla all over again, but now with a medieval baptistry thrown in. Here all was cold marble. There was the ceaseless trickle of water over stone. One's voice grew resonant. Facing the entrance was a vast urinal into which a boy of nine could have stepped and been accommodated. Once I did just that. The thing was five and one half feet tall, two and one half feet wide and partially recessed into the wall. A marble plinth extended in front of the urinal upon which one stepped to void. This gave a certain elevation to the proceedings. One rose to the occasion. Positioned for duty, one was concealed from one's neighbors by projecting lateral aprons of stone which met above in a kind of oval nave. In such a grotto a saint might have stood to receive veneration. The marble itself was etched with a plexus of interlacing lines and cracks such as is seen in old oil paintings. It was as though something lay just beneath the stone surface which, after a long unblinking stare, might emerge. A scene, perhaps, of the Battle of the *Monitor* and the *Merrimac*, or *The Rape of Lucrece*. At the bottom of the niche was a wire screen upon which had been placed a votive offering of raspberry-scented soap, always in varying degrees of erosion as the God of this place absorbed his due.

Into such a receptacle, urination was performed importantly; it was grandly addressed. One felt larger. One had significance. Class distinctions were abolished. In this place, chairman and clerk, paper boy and publisher, all occupied the same station in life. Before the Grand Urinal of the Elks, all men were brothers. Nor have I met with so democratic a plumbery in forty years of urinary experience.

The sacristan of the shrine was an elderly man who, wearing a fitted waiter's vest, performed his tasks with the kind of serenity that comes over those who tend altars. He had a scrubbed, luminous face which had taken on the color and glow of the marble amid which he dwelt; I should not have been surprised had it the same cool, smooth texture. It was Ferdinand who extracted with tongs the cigarette butts trapped in the strainer, replenished the raspberry soap-cake and sifted the sand in the urn for wads of chewing gum and other rejectimenta. I see him now, a freshly laundered towel over one arm, polishing the brass of the faucets, wiping and rewiping the gilt-edged mirror over the sink, and all in absolute silence, as though the god Harpocrates, with finger at lips, had shushed him. It was Ferdy's particular genius to see everything, but never seem to. In Ferdy, men's room attendancy was exalted to a high art.

What a far cry, all that, from the modern toilet wherein a man, in haste, as though furtively, and certainly joylessly, does what is to be done and no more, sending his liquefactions swirling down a bowl sent vortical by the depression of a handle. How puny by comparison is the dwarfed and distant urinal of the present day, which delights not the senses nor spurs the imagination, and where one is expected to aim to Muzak. Even the substitution of a picture of a top-hatted man in profile for the word *lavatory* is symptomatic. Urination has become cute. And if that can happen, can the fall of civilization be far behind?

Recently I returned to Troy and to that Elks Club which I had not entered for four decades. There, under the influence of certain celebratory liquids, I heard, at first faintly, then with more and more urgency, the half-remembered call of the Lavatory. It was right where I had left it, just off the lobby. There was the same ancient sign above the midriff swinging door. I pushed through and found myself standing where I had stood so long ago, listening to the sympathetic trickle of water over stone. I looked about for Ferdy. But he was not there, of course, having long since been assumed to glory. In a magical daze, I mounted the single step before the great urinal. I arranged myself for voiding, relishing all the while the sweet sense of enclosure, the intoxication of grandeur. There I waited for the kind of surcease I had not known in forty years and for which I ached with all my heart. But it was not to be.

I should have known better. One cannot recapture the storied past. Wait as I might, try as I might, cajole, reason, curse, as I might, I could not. Nothing. Only a few pathetic drops. All at once the Grand Urinal of the Elks took on the shape of a man-sized coffin, its marble maw open and mocking. For one menacing moment, I feared that the thing would clamp down upon me with a crash of colliding crockery. I would be cut off, chewed, mangled!

With bladder bursting, I turned and fled, scarcely attending to zipper and button. Through the swinging door, across the lobby and out to the street. Then around to the back of the Elks Club, where against the bark of a kindly maple tree, I let go a thick twine that would have been the envy of Hector, Priam, Paris or any of the old gang of Trojans of days gone by.

Feet

———— ✦✦ ————

At one time, his feet were what a man worried about most. What with thorns, slivers, bone bruises, scorpions, asps and chiggers. Feet carried you to the hunt or the harvest, brought you to the defense of your village or let you run away. Time was, when you lost a foot the rest of you was not far behind. Oh, you might drag and hop for a while, but finally you just sat down and waited to die. Although it may no longer be true in this time of blood-vessel grafts and prostheses, it is still truer for the aged than it is for the young. The aged, for whom two working feet are the only hope of giving oneself an airing. Had I to choose any single specialist to tend the elderly, I would choose a Chiropodist. He strides among the aged like a god among mortals. On the other hand, there is courage and laughter. There is a woman with diabetes whose leg I amputated. Once, she had been a Latin teacher.

"What do you think?" I asked her. "Can you learn to walk with an artificial leg?"

"I can manage," she said.

"You plan to walk on your hands, then?"

"What do you mean?"

"Manage," I explained. "From *manus, manus*—hand."

Her laughter informed me that she would walk. (Keats had it wrong. In the very temple of despair, veil'd courage has his sovereign shrine.) Six months later this Latin teacher walked easily into my office for a checkup. When she left, there was a brown paper bag on my desk. It had the shape of a specimen bottle, but the shape only. For this bottle held her own homemade slivovitz, a concoction of pure grain alcohol with a sediment of plums. "Eat these plums," the attached note said. "You look peaked." I did eat them. Every one, and called to thank her for the lovely gift.

"It wasn't the plums," she said. "It was the spirit in which they were given."

The smartest thing a surgeon can do is to wear comfortable shoes in the operating room. I used to operate in a pair of shoes from which, for comfort, I had cut away the tops of the toes. *Qua* footwear, they had not their equal in the world. I had always thought them my lucky shoes until three days ago when, crazed, I suppose, by the tedious pickery of an Assistant Resident I was guiding through a gallbladder operation, a deep Deaver retractor leaped from the incision and threw itself upon my foot. Stifling a howl, I kept on with my pedagogical martyrdom until the rock-filled gallbladder lay smouldering in a basin. Only then did I bend to remove the shoe and sock from my foot to discover that, while the four smaller toes sprouted gracefully as before, Hallux had burst into sudden purple bloom. He was the size and color of a plum. The toenail had become elevated from its bed by a collection of blood for which there was no mode of egress. Of the pain of subungual hematoma I shall say no more.

Even in his agony a teacher's first thought is of his sacred charge. Calling for a paper clip and a book of matches, I summoned my students to the locker room. There, I straightened out the paper clip and held one end of it in the flame of a match until it glowed. I pressed the red-hot point against the base of my toenail just distal to the cuticle until the metal cooled. There was the odor of burning toenail. Again and again I heated the clip and applied it to that spot until, at last, there was a sudden *give*. I yowled and pulled the clip away. It was chased by a gush of black blood from beneath the nail. Suddenly there was no pain. I smiled up at the adoring circle of students.

"You see?" I purred. "So much for your vaunted technology when a paper clip and a match will do."

That night, the pain was marvelous. My toe had swollen to the size of a peony into which the edges of the nail cut cruelly. In my zeal to instruct I had leaped from the narrow ledge of subungual hematoma to the abyss of unguis incarnatus septicus. Soon, I knew, there would be gangrene. Then amputation. For three nights, from midnight to dawn, I listened to my toe singing to me with the voice of David outside the tent of Saul. For three days and three nights, I soaked my foot in warm saline and smeared it with unguents, now and then rising to hobble on Rounds, casting prayers into every corner of the hospital, this unrepentant mass of bricks and mortar that is so much larger inside than it is outside.

Shall I never again keep both feet upon the ground? I wondered. Shall the weight of my body no more be tossed with careless grace from leg to leg? No more left, right, left? And, as is the wont of all mankind in misery, I set blame upon my poor mother. For I surmised that she had long ago dipped me in a sacred pool holding me only by my left big toe. Constitutionally unable to perform rituals as specified, the dear soul had misread the directions. The very touch that was to protect me had rendered me vulnerable.

It was three days later, and I was saved. The peony had shrunk to plum and then to Concord grape. The pain had become a distant thunder. I forgave my mother, and in the joy of recovery I contemplated upon the matter quite differently. That state of one-leggedness toward which I had been "retracted" and from which I had now been retracted, seemed to me a noble State, whose citizenry is marked by lameness for a special purpose. They are Gods and Heroes who dwell there. Had I been given back my leg at too great a cost? I thought of the others of this lame tribe—Hephaistos, the Limper, the only one of the gods to accomplish anything useful; he makes swords and shields and jewelry while the others loll about Olympus being lewd. And Bellerophon, whose limp was translated into the wings of Pegasus. And Ulysses, gored in the thigh by the tusk of a wild boar. And Oedipus, and one-sandaled Jason and all the rest. Perhaps, had one-footedness been thrust upon me, I, too, should have entered that elect band of heroes, more likely to teeter and fall, I grant you, but marked for greatness, greatly wounded, and moved to create Art as my wooden leg.

This letter has been interrupted by one of your colleagues, a medical student who wants to become a writer. He asked me how it was that, after twenty years as a surgeon, I began to write. It turns out that the answer to that question has to do with feet. In the interest of unity, I append it here.

A surgeon is not unlike a shoemaker who lives in a state of peaceful dreaming. The shoemaker loves his work, the feel of his bench, the smell of leather, his knives, the knock of his hammer. These things are companionable. For them, he feels a deep courtesy. In their presence his spirit is lulled, relaxed. He sighs with contentment. And the emergence of each new shoe from the pad of leather is for him a small miracle that he has somehow wrought. His days flow on, accruing to compose his life.

Then, one day, he is distracted by something—the honking of geese from high above the clouds, an odor of lilac or hay, a sound or smell that brings with it a long-forgotten moment. He looks away from his bench, daydreaming, entranced. When he looks back, his knife, his awl, his hammer, seem strange to him. True, they ride his hand with the same confidence, but now he is keenly aware of these tools that he had taken for granted as extensions of his body, and of his bench, and the muscles of his naked forearm as he grips and bends the leather. Where before he had felt only the rhythm of his work, now he has become conscious of himself. Perhaps he has lost his childlike innocence. Perhaps he has begun the process of dying.

One day I saw a winter meadow covered with a smooth expanse of snow. Glacial, still, perfect. The next day I returned to see that a line of footprints had advanced from the edge of the meadow toward the middle. Peer as I might, I did not see them complete the traverse to the opposite side. Perhaps it was too far to see? But no, I walked around the circumference to see if they had come across. They had not! Nor had they returned. Perhaps . . . perhaps they stopped in the middle. But how *can* that be? Now is a time of exaltation, for these footprints prick the imagination as no feet ever could.

They have introduced a new element, the element of narrative, history, character, plot. Was the creature who made them . . . winged? Did he at that point where the tracks leave off, did he therefrom ascend? Take wing? Preposterous! Surely, the footprints continue on to the other side of the meadow. Were I to follow them with my own feet, stepping carefully into each depression, or, more respectfully, alongside, surely then I would pick up the trail. Something keeps me from it. Is it a reluctance to further wound the white body of snow? Is such a defilement too high a price to pay for knowledge?

Perhaps I do not really want to solve the mystery. To solve it would place a limitation upon these marks. I would know exactly from where to where they go. I might measure their size down to the millimeter. I could deduce the rate of speed at which they were made. I could infer their gender. But I shall not. Better to depart and return once more to scatter sweetened bread crumbs on the snow to feed the angel should he come back.

And so I have gone from feet to footprints. Soon, I suppose, it will be footbeats, those sounds that hover between footprints and feet. Footbeats. You reach out for them the way a child reaches out to capture a shadow. He opens his fist and . . . nothing.

Feet

Twenty-five years ago, I sat upon a high stool in an Anatomy laboratory in Albany, New York. It was a lofty stone vault at the very top of an ancient building. Through skylights placed here and there, pearly shafts of light slanted to the dissecting tables. To step through these shafts of light was to undergo a mythic transformation, for the motes and specks one saw therein were nothing but pulverized floaters, a holy drift from the cadavers themselves. Take a breath, and you inhaled them; stir, and they settled on your hair, your skin. It was a kind of anatomical intimacy with the cadavers. To me, they seemed so stately, those dead, enduring their daily measure of shredding, pulling, gouging. Before long, they were meat wrecks—something made of leather, something unfinished—a strange collage of wallets, pouches and lanyards. It is just and sensible that in order to become doctors we must dip into the very dead to lift forth our birthright. It is only by drawing upon ourselves like sleeves the bodies of the dead that we may come to it, this knowledge that is ours and no one else's and that must be committed only to the most worthy.

I am like the shoemaker who has sat for a lifetime at his last only to discover in one moment of revelation that he is preoccupied with footprints rather than feet. Feet, he knows, are a concretion, a fact. Footprints are a mystery; they have an unknown past and future. Feet declare, footprints suggest. Yet it is often the footprint which survives the foot. The footprints of prehistoric man in the ashes of the Olduvai Gorge have remained, reawakening the errand of the feet that made them. From what longing departing? Toward what joy approaching?

Only recently has man lost his fear and reverence for footprints. Before, footprints had magic powers. It was by his footprints that a man was undone. First, discovered, then pursued, then slain. Injure footprints, and you injure the one who made them. A man who had special reason to dread the spite of an enemy had to efface his footprints with a branch as fast as he made them. I like the idea of sympathetic magic; it indicates faith in the oneness of nature.

And footprints revealed the passage of wrongdoers. Nine hundred years after his eviction from the Garden of Eden, Adam lay dying. He ordered his third son, Seth, to go to Eden to beg for a drop of the Oil of Mercy so that he might not die. Seth would know the way, his father told him. He had merely to follow the footprints Adam and Eve made when they left Paradise long ago. For wherever the first sinners stepped, the ground was burned and no vegetation was ever to grow.

It was in the anatomy laboratory that the poet John Keats called a halt to his medical education. Urged to attend medical school by his father, Keats proved a desultory student. One day, while seated at his cadaver, a sunbeam strayed through a skylight and fell upon him where he sat. Keats glanced up and saw within that beam of light a host of elves and fairies dancing. All at once he rose from his stool, mounted that sunbeam and ascended to the skylight. Then through the glass, and into the open air. Keats never once glanced back. For him, it was the end of Medicine and the beginning of Poetry. Now, with one leg in each of these disciplines, I can report that both are subcelestial arts; the angels disdain to do either one. If forced to choose one way or the other, I would suffer. But in the end, I would fall to the medical side, agreeing with Walt Whitman, for whom the sight of the wounds of a Civil War soldier "burst the petty bonds of Art."

Impostor

There is a certain region of Asia, not far from the Arctic Circle, where the villages are scanty and remote, and where the customs and pace have not changed for three hundred years. Because of the poverty of the soil, the landscape, the stinginess of the sun, the wolves, for whatever reason, the rest of the human race has not rushed to visit this place. Even Genghis Khan had given it a wide berth. There were czars who had never heard of it. The sole occupation of the people is the felling of trees for timber. But if there is no wealth, neither are there any beggars. Their only nod in the direction of government is the existence of a village chief, who is the eldest son of the oldest man in each village, a person of middle years and with a venerable heritage, for old age is venerated here. The only recreation through the long glacial winter is to study the sky and to repeat the tales of their ancestors. Now, only bureaucrats come to take the census, levy taxes and to enforce regulations. And once a year, the Health Inspector.

Even after N. had been in the village for a year, it still amazed him that no one here had ever beheld the sea. It made the inhabitants seem drier, heavier, more solid than he. Not less, only different. The way a sturdy black clot differs from the coursing liquid form of blood. In the absence of a sea, the people hunkered and stared up at the sky for hours at a time. It seemed to satisfy them the way the sea satisfies those who dwell upon its shores. Perhaps, he thought, there was once, eons ago, a sea that covered this place, and this village then lay on the edge of a great ocean which long ago had turned to vapor and been gathered into the sky, and which now the people studied as though it were the sea. It was a village without church or priest, a pagan place that had gotten separated over the centuries from its Christianity. In the absence of faith, he thought, superstition did them very well.

The time of his coming was still what passes for autumn in these parts, and the woods were in waning leaf. Just at sunset he broke from the matted forest upon a field that had been planted with potatoes. He paused there, legs apart and wobbling. The sound of his breath scraped at his ears. Shielding his eyes with one hand, he gazed beyond the field at the village which glowed and undulated like a molten core. Was it real, or another hallucinated place? He had seen so many mirages. Near the far side of the field two men squatted, working potatoes out of the ground. Had they glanced up at that moment, to one he might have resembled a newborn colt making its first boneless stand. To the other, he could have been a wild creature caught in the last agonal lurch of its life. But they did not look up to see him.

With his coat flying wildly about his legs, N. staggered across the open country and into the streets of the village itself. Before long he came upon a small, cobblestoned square from which the rest of the town radiated. In the middle of the square was a stone horse trough, or so it seemed to the dazed and delirious stranger. He bent over the trough. Again and again, he scooped water and dashed it into his face, anointing his head, ladling it into his upturned mouth. It was sunset, and the empty square was heated to a deep red. The water in the trough too was red and thick as temple oil. N. straightened and stared at the sun, which seemed to him a coin that had been fixed and minted at that spot. All at once a crow flew across the sun at its very middle, bisecting it with its flight, then emerged into the honeyed sky. N. felt a sudden sharp pain in his temple, and something else—a fluttering as though the flight of the bird had entered his own blood. An irresistible giddiness came over him, and he fell unconscious to the cobblestones. But not before he knew that he had found his place, that here, in this tiny square, he would end his flight.

N. awoke foaming and with all of his muscles having brawled among themselves. He was in a bed. A woman was standing over him. At his first stirring, the woman leaned nearer. He saw that she had a harelip that completely bisected her upper lip to the nose. Through the rent he could see the moist red darkness of her mouth. It was like a fresh wound. Because of this, and because she was wearing a kerchief around her head, he could not judge her age.

"How long have I been here?" He wanted only to know how long he had been at the mercy of strangers. He did not think to ask where he was.

"Three days." The voice sloshed in and out of her nose.

"What is your name?" he asked.

"Mona." She began to question him.

"Where did you come from? What are you doing here?"

He closed his eyes. After a long silence he spoke. "I am a doctor."

From the finality of his voice she knew he would tell her nothing more. He took the soup she fed him from a spoon, with the obedience of a child. When he had finished, the woman left the room, and he slept again.

Heavy bootsteps on the wooden floor awakened him. Again he saw the woman and a heavyset, bearded man wearing a fur cap.

"Ah, so you are awake. Good. I am Seth, the chief of this village. Not that a chief is needed here. Like the others, I work in the woods." He fell silent as though there was nothing further to say. At last he spoke.

"You are a doctor." He stated this as a fact that he had always known. "We have had no doctor here for as long as anyone can remember. Many of the people are sick. The phlegm in the streets is red. The people say that you have been sent to them. They promise that if you stay here they will build a cottage with a table and chairs and a bed. There will be a garden. This woman will serve you . . ."

N. was silent. The man continued. . . .

"You must remain, anyway, for the winter. There is nowhere to go. Soon it will begin to snow. It is beginning already, and the road will be impassable."

"I will stay," said N. "Only . . ."

Seth waited. "Yes? What is it?"

"I do not wish to answer any questions. Nothing. I am . . . who I am. That is all."

Seth nodded. "I will tell the others."

Alone with the woman, N. gazed up at her in silence. Mona dropped her glance from him as though to hide something which might have been revealed there. The truth was that N.'s coming had been foretold to Mona in a dream. In her dream, she was a child again and had been sent by her father on an errand. She did not know what it was she had been sent for, only that she must retrace a trail of footprints in the forest where no vegetation grew, and where the grass had been burnt by the steps of whoever had walked there. The path led to a clearing at the center of the woods, a place as bright as childhood. Here the light poured through the trees from a source high above. In her dream, Mona approached the bright place as though beckoned. She held out her hands for whatever she was to receive. All at once the light softened and dimmed. Mona opened her eyes. In her hand was a single drop of pearly fluid.

When she awoke, Mona felt reassured in some deep sense. She dressed and ran to the fountain in the square, understanding only that the time and place of the rendezvous mattered so. From the edge of the square she had watched him tremble and fall. In the few seconds it took her to reach his body, the sun dropped from the sky like a stone.

From the beginning, the people came to his cottage seeking relief from their pain and cough, and for the repair of their wounds. Early each morning, Mona came, bringing food and clean linen. All day she worked with him among the patients. He taught her which herbs to gather, and how to grind them in a mortar and pestle, how to measure out and mix the ingredients of a medicine, how to bathe and dress a wound, how to care for the instruments that he whittled and carved from the bones of squirrels and mice. Each new task she accepted eagerly, but with virtually no speech. She had learned early not to inflict her nasal wheeze on others. Almost at once, N. seemed to know the villagers in some profound and ancient way; not their names as yet, nor their temperaments, nor the facts of their small histories, but their bodies, both individually and collectively. At times, he thought of the village as a single body that had been given him to tend, so that when, here and there, a part of it fractured or burst or wilted, he would turn it over and over in his hands until the damaged spot became visible to him, and he would repair it. In the weeks that followed, what had seemed a village of freaks into which he had stumbled became a village of friends to whom he was bound by a thousand cords of trust. These people would risk everything to save him; he saw it in the faint smiles they let play upon him when he passed.

To the villagers, N.'s diagnoses and treatments were infallible. What he said was the matter with someone became the matter with someone. As though all disease hastened to mold itself to corroborate his intuitions. To bathe the wounds of his patients, N. used freshly melted snow from the center of virgin drifts. The knives he chipped out of slate were sharp and gray as the edge of a pigeon's wing. Once, to suture the thigh of a man who had been clawed by a bear, he used Mona's long black hair, twisted and waxed. There was a woman whose husband had been covered with boils. Before her eyes, he cut open each one of the abscesses and expressed the pus. Then he applied an unguent on a cloth. A week later, the wounds were healed. Nor was there later any sign on his body where they had been. Later, the woman would tell how he had plucked the red sores from her husband's body one by one as though they were flowers. Often, he would coax a patient into sleep, and heal him through dreaming. Before long, the village was like his clothing that he wore next to his skin. Here and there, there would be a tear in his shirt, or the stain of berries. Then he would wash and mend the injured shirt, and put it back on as before.

He had trained himself to gaze at his patients as though he were Adam and each patient were the second human being created. One morning from the doorway of his cottage N. watched three men emerge from the forest. They came across the open field and headed for the village. An old man walked between the two younger men. The old man was bent forward and to the right, as though pulled in that direction by an invisible rope. With his left hand, he gripped his right wrist, supporting it at a fixed distance from his body. The right arm was not allowed to move. The old man's legs were not injured. His step was sure. His head, too, was carried without pain. The men entered the cottage and told their story. Old Rolf had lost his footing on a muddy slope, and had fallen on the outstretched arm with which he had meant to break his fall. There had been a sudden pain in his right shoulder. He was unable to move it.

N. took the right wrist in his own hand, supporting it at the same angle, and he helped remove the old man's jacket and shirt. Then he gave back the injured arm to Rolf, who held it once more, rigid, listing. Now N. could see the difference in the two shoulders, the right being hollow above, and with a lower bulge where the head of the bone had slipped out of its socket. N. motioned Rolf to lie down upon the table. The old man did so, never letting go of his wrist. Now N. knelt to remove his own right boot and sock. Once again he took the wrist of the injured arm in his own hands. This Rolf relinquished in silence. N. raised his leg and gently inserted his bare heel into the old man's armpit. Bracing the body of the man with his heel, he drew down on the arm at the same time, turning it outward. Gradually he increased his pull, until there was a sudden muffled sound as though an apple had fallen from a tree into wet earth. At the sound, the two men, Rolf and N., looked at each other and smiled. N. moved the shoulder through a gentle range of motion, then returned it carefully to the side of the old man. Weeks later, he came upon Rolf chopping wood in front of his cottage.

N. seemed always to be available to the villagers. They had only to seek him out, and with rare exceptions, his whereabouts were predictable. Anyone who was injured or ill knew exactly where to find him. When he was not needed, N.'s existence was not obvious in the village. He faded from the consciousness of the people, remaining as a vague continued presence among them. Only the loss of him from their midst could have informed them of the extent of their calamity.

Far from resenting the indifference of the villagers, N. enjoyed the solitude afforded him to fashion his instruments and create his medicines. Sometimes, it seemed, he cultivated his strangeness. He was not one of them, nor did he wish to be. Only to dwell in the shadow of the village, ready to be summoned or withdraw. Now and then in the evening Seth came to sit with him, bringing a wooden figure that he would set about carving, while N. sharpened his own tools. Mostly, there was no talk, only the silent working in each other's company. Each visit was closed with a glass of brandy. Gradually N. came to count upon Seth's visits, his quiet companionship. Once, when Seth had remained away in the forest for days, N. felt something akin to loneliness.

To the rest, N. was nameless as light. Why had he come there? How? From the very beginning, they assumed that he had *fallen*, in some mysterious way, and they received him with the largeheartedness of people who suspect that someone has already suffered quite enough. Still, they wondered what it was he had done. One old woman, with the audacity of the aged, asked him point-blank what crime had driven him to this "Godforsaken place." Had it to do with a woman? He had been sanding down the calluses on the old woman's feet with a rough stone. He smiled, and without pausing at his task, he told her with mock seriousness that he regretted only those sins that he had not had the time or strength to commit. The old woman shrieked and cackled. Mona was outraged at the woman's familiarity. She clamped her hand over the toothless old mouth, and whispered something fierce into her eyes, after which the old woman gasped and sat perfectly still. But N. knew that these people would have forgiven him everything. Murder, even. Such an act could only have been a momentary lapse, not anything to be weighed against him. And without doubt, it would have been justified. If someone should suggest that N. was secretive, someone else would counter by saying that he didn't like wasting his breath. When he looked back at his fugitive months and years, N. was unable to quite remember the pain in his lungs, nor the thorny feet upon which he tottered from city to city, from forest to cave, nor what sustained him, what food, what drink. Only that he had fled and escaped to this place, at this time, and that, without any guidance or help, he had kept a rendezvous.

Every morning at dawn N. returned to the trough in the square to bathe and drink and feel the reflected moon take shelter in his body. The trough itself was of stone and made like the plainest, most ancient fountain. It was bare of ornament. Only that beasts be led there to drink. But surely, he thought, this was more than mere trough which holds standing, unfed water. This water flowed! It refreshed itself. The more N. gazed into its depths, the more he saw that the trough was, in fact, a small bridge of stone spanning a stream that ran beneath the village. Only a small portion of the running water was led into the catch basin which was the trough. The rest flowed on as a heavy tide. Whatever dark cold energy he drew from the stream was reconstituted into something he could use to heal himself and others. He was learning ancient laws. This stream that he tapped had been flowing since the dawn of mankind. Not even the sound of the cobblestones crying out: Move on! Move on! Not even that warning could have broken his fated sense of homecoming.

In this manner three years went by. One evening, after serving him his meal, Mona tidied up and made as if to leave for the night. But she did not go. Instead, she paused at the table where he still sat, and stood above him in silence. He understood that there was something that she wanted to say. He looked up at her, questioning. At last she spoke.

"My lip. I want you to fix my lip." The suddenness of her request shocked him. All at once, he was trembling.

"That I cannot do," he said quietly. "It is too difficult for me."

"I need you to do it." Her voice, even with the leaking air, was precise and strong. The clarity of it surprised him.

"No. I cannot do it." N. could not have said why. He only knew that he could not. All at once her manner changed. Her voice rose.

"Why? Why? You must help me. You must!" She turned her face and wept into her hair until it was heavy with tears. The breath tore up out of her throat. She doubled over in anguish. He watched the heaving up and the swallowing back down of her disappointment, how it gave and tightened and gave again in her chest. All at once, N. saw that it had been a hope that she had held close to her heart from the moment he had arrived. And hope was something she had never felt before. Had she not attached herself to him like a serf? He waited for her sobbing to subside.

"It is just a small tear," she whispered, "a torn piece."

N. made no answer. Again she was weeping, like a child, with no restraint, as though a whole part of her mouth had been eaten away. N. sat at the table, his head inclined away from her, fending off her sounds. All at once, she fell upon him in fury, punching his head and shoulders and chest with her powerful fists. Seizing him by both arms, she shook him as though to shake out the sutures that she needed, that *must* be lying there like coins in his pocket. What she could not tell him, speaking with her fists, was that there was no way to hide it, that only the night covered her shame, that she had lived like a bat until he, N., had come and led her forth into the light of day. It was he who had done this to her, given her hope. And now, he must. He must!

For a few minutes, N. submitted, then rose to his feet. He put his arms around her, steadying and imprisoning her. Suddenly, she was still, like an animal that slumps into torpor in the jaws of its predator so as not to feel the killing. Carefully and with great precision, the way one fits together the two parts of a broken dish, N. covered her mouth with his own, so that the harelip of the woman was held between his own lips. To the woman it was like a branding. His refusal! And now this! She struggled to free herself, rearing backward, whipping her head from side to side. Her nails dug and raked his back. She bit his lips. She felt herself dying. Blood was coming through his shirt where she had raked him. She tasted his blood in her mouth. In their struggle, the table skidded, a chair overturned. And still he held her. Then, with a quick movement, he bent her backward to the floor. He lay on top of her, pressing against her soft belly between waving, frantic legs. Again he placed his mouth on hers. His tongue was a wasp, stinging the edges of her cleft lip first one way, then the other. Exhausted, she gasped into his open mouth, then pulled his breath back into her lungs, and felt the warmth of it inflating her body, floating it . . .

Holding her pinned, the ripped and twisted face so close to his own, for a moment N. himself was afraid. The ferocity turned her features into the very jungle through which he had fled for so long. But now N., too, let the kiss take over, submitting to it, he as well as she, stoppering their mouths, stopping up her anguish and his helplessness, sensing somehow that the deed would speak the truth for them both to hear. And he, for his part, when he sank his hands into her long and powerful hair, encountered a braid so thick it satisfied his whole fist. And he, for his part, drank from her as though he were a sunbaked rock at the edge of the sea, and she were the cool foam of the incoming tide all trimmed with bubbles. Now his tongue was like a cunning knife; her mouth was bathed in his medicinal saliva. Mona felt her lips heating, growing hotter and hotter, until her whole mouth turned molten and fluid. There was a flowing, a fusion. Something, a process that had been interrupted long ago, was now being completed. She had a moment of rapture in which healing and ecstasy were indistinguishable.

For a long time they lay still, as though sleeping. At last N. disengaged himself from her and rose. He gazed down at her and his eyes filled with tears. Before he had seen her as whole; now she was wounded in a way that no man could heal. Poor woman, he thought, now you have caught the wound you have longed for all your life. From that day, Mona lived with him in the cottage.

It was spring when the Inspector arrived from the capital. It had been a long journey, first by riverboat, then on horseback. There had been no warning in the village of his visit. Only, perhaps, a slight graying of the sky, which might have been enough to warn the older people, but this year it did not. The Inspector was a short, thick man wearing a long black coat. With the seal of office around his neck, he could have been mistaken for nothing else. About his mouth there lay the lax fatigue of an overtired child.

The Inspector seated himself at the table in N.'s cottage, and took a pack of playing cards out of his coat pocket. He shuffled the cards elaborately and for a long time, then began to deal them out. Each card was served with a small, sharp slap upon the table. Only when the game had been fully set did he begin to speak.

"It seems to me that we have met before?"

"That is always possible." N. was trembling, but willed the muscles of his face to freeze.

"Have you served in the military?"

"I do not like to speak of it."

"Were we together for a time in a hospital during the Southern campaign? I think we were. It seems that something happened. I cannot recall it for the moment. Never mind. I shall, in time." The Inspector had the same smile as a scarecrow, and of the same sincerity.

"May I see your credentials? It is a mere formality, you understand." N. made no answer. The Inspector gathered in the cards and riffled them once loudly. "Your diploma, your license to practice medicine."

"I have no credentials."

"What! Nothing to prove that you are what you say you are?"

"I have no papers."

"But this is impossible! What if I say that I do not believe that you are a doctor. Perhaps you are an impostor." The scarecrow smiled again. "Don't take offense, please. One has a right to ask, hasn't one? After all, the people are not sheep to be led to pasture each day by some mysterious shepherd. Come, come, I must have your credentials."

"I have no papers. Doctoring is my work." The words were pronounced quietly and with no trace of outward defiance. He was merely offering an explanation to someone unfamiliar with the customs of a foreign land. Above all, N. wanted to explain to this man that it was not diagnosing that he did but deciphering. He was, it was true, no doctor. He had merely been given the secret of a code.

The Inspector rose from his chair so abruptly that the table slid and one of the cards fell to the floor at his feet. N. knelt to pick it up, then held the card up to the Inspector who towered over him. N.'s outstretched arm and upturned face were suppliant. At that moment, N. felt no fear. Only the sudden immense and painful onset of old age.

"It is of no use," said the Inspector. "I remember everything. Who you are, what you have done. An epileptic! How is it that you presume to cure others when you have not the least idea how to help yourself? But this doesn't matter any longer. You are under arrest. You are to remain here under guard until transportation to the capital can be arranged."

In the house of Seth, the Inspector did not smile. "That man, your 'doctor,' is nonesuch. He is a fake, a liar, a criminal. He is wanted for murder."

Even at that moment, word was spreading from mouth to mouth. The village took on the nature of a town that has been seized and occupied by an enemy. The few survivors of the siege were now being interrogated. They would like to resist; they did not know how. Seth sucked in his saliva.

"How do you know . . . ?"

"I knew him at once, of course. We were stationed at the same military hospital fifteen years ago. In the south. I was in charge of medical supplies; he was an orderly in the surgical wing. It was a famous case. Even then he was a stranger, kept himself apart from the others. Efficient, I'll grant you that. And dependable, or so it was thought until the truth came out. This man is an epileptic. He is subject to violent fits and fits of violence." In answer to Seth's astonishment, the Inspector went on. "So! He has managed to conceal it here also. He is clever enough about it, knows in plenty of time when a seizure is coming on, and he keeps out of sight. It all came out in the court-martial."

"But epilepsy . . . it's no crime."

"One day he had a seizure without any warning. It took place in the middle of a ward full of freshly wounded soldiers. For the first time his convulsion was witnessed. Just before he fell unconscious to the floor of the ward, he leaped upon one of the injured soldiers and strangled him. The man died. When he awoke, your 'doctor' professed no memory of the event, of course. It was later learned that the dead soldier had lent him money. Your man was arrested. A court-martial convicted him of the murder. He was sentenced to be executed by the firing squad. On the eve of his execution, the man broke from his guard and disappeared into the jungle. He was never seen after that." The Inspector bit off the end of a cigar and spat it into his palm. "They tracked him with dogs. Each time they came close, he would slip away. But now . . ." He gave a tiny smile.

"What will happen to him?"

"I leave for the capital in an hour. You are to arrest and guard him day and night until the police arrive to take him off your hands.

It will be a matter of six or seven days. If he is not here when they arrive, this entire village will be rounded up and treated as prisoners. I promise you that."

"Arrest! Guard! As though he were a thief! Listen, Excellency . . ."

The Inspector had turned to go.

Seth ran to catch up with the striding man, calling out to him, "What are certificates to us? For three years he has lived among us. He had brought us nothing but comfort. He is a good man. And so many have been cured! Old Rolf, the Abban childbirth . . . you have heard the stories yourself."

"Coincidences. I am not ignorant of these matters. Come now. The truth is that most things will cure themselves if left alone by doctors. The less meddling, the better. The human body knows how to take care of itself. You insist that this man meant well? So be it. Motives are not my concern. Credentials are." The Inspector had already untethered his horse, and mounted. From the height of the saddle, his voice took on a warmer tone. "I am interested in the truth. I do not settle for illusion, no matter how sweet. Why, I believe you people don't care a bit that he is in fact an impostor."

Seth did not answer, but glanced quickly up and down the street as though to commit some act that must not be witnessed. But he made no move. The Inspector smiled.

"Tell me something. Why are you so intent upon protecting him?" His voice was quieter now, insinuating, almost musical.

Beneath the cold smile of the Inspector, Seth shivered.

"Because . . . I love him." But his words never carried to the Inspector over the quick sound of the hoofbeats.

Once again the two men sat at the table in N.'s cottage, as they had so many evenings for three years. Seth spoke softly, behind his hands as though fearing to be heard.

"Why did you do it? Why did you lie and pretend? I must know."

"It was what I expected of myself."

"Weren't you afraid to fail?"

"There are worse terrors. I knew that I could do this work, would not shrink from it. Nor did I imagine it very difficult. And I was right. It is in largest part a matter of craft and intuition. I had only to make and use objects, and to retain the clarity of childhood."

"The people will not accept it. We will never submit. We will rise up."

N. shook his head. "No. You must not resist. In the end it would all be the same."

"But this time you are wrong. There is the small matter of the human spirit."

"No," said N. "I forbid it."

Seth rose from his seat, and came around the table to face N., who had remained seated. In a moment, Seth was on his knees looking up at N. "Please," he said. "We will find a way. What is different today from yesterday?"

"Everything is different."

"But we are all impostors of one sort or another. The worst thing is not to know you are one. That is the real pity. And falseness can collapse under the weight of good work. You have shown that. Please." Seth threw his arms around the seated man's knees, pinning him firmly, as if at that very moment he intended to walk away.

From the beginning Mona had known of his seizures. On that first day, had she not watched him awaken chewing his saliva? Had she not, on that first day, examined his body, looking for inscriptions on it? Once, when they had gone into the forest together to gather herbs for medicine, they had become separated for a bit. All at once she had heard a terrible cry like the scream of a parrot. She had run then in the direction of the sound, calling out for him. She had come upon him lying on the ground in a deep, unrousable sleep. All about him, the stalks of the underbrush had been trampled and bent as in a violent struggle. She had decided then never to speak of it.

Nor did she speak of it now. Such a thing was private, unsharable.

N. could tell, often days ahead of time, that an attack was coming. Evidence would begin to accumulate in his body. Feelings of strangeness, dreams, a sudden jerking of his left leg, palpitations, giddiness, unrestrainable laughter, a bunched-up pressure. Above all, that trembling in the pit of his stomach, as though his blood contained the flight of a bird. When the time came near, there were other signs—flashes of light, a hissing as of a great serpent. Long ago, as a youth, he had tried to stifle that hissing with his hands pressed over his ears. But it was of no use. Whatever it was that made that noise lay coiled somewhere within him. At the last, there was a jolting blow to the side of the head that knocked the world aslant. Then, nothing.

Always, on the day after a seizure, he was able to see more clearly than he had before. As though the convulsion had somehow blown off a haze that had settled across his mind. At these times, he knew a rapt penetration into things.

From the day of the Inspector's appearance, N. had discontinued his morning visits to the trough in the square. The accusation of the Inspector seemed to forbid it. The very words of the man had rendered him unworthy. Even in his own mind, N. felt himself humiliated. For the first time, his door was barred to the villagers. Hidden, and alone except for Mona, he awaited the police. On the third day, there was a knocking at the door. Mona opened it to Seth, who entered carrying in his arms the body of a small boy, about ten years of age. A glance informed Mona that the child was dying. She motioned Seth to lay the child on the cot. She waited for the man to leave, then closed the door after him.

"It is a child from the next village," she told N. "The father has been walking for two days to bring him here. His name is Mihail."

N. rose from the chair with great effort. He seemed ancient, ossified. His movements were ponderous. He stood by the side of the cot and reached for the boy's pulse. He found it and trapped it between his fingers. What he felt there was no more than a twine of falling water that wavers in and out of existence. Each time he thought he had it, the pulse darted away. The eyes of the boy were sunken and glazed. They roved aimlessly, now and then disappearing behind the upper lids. The thin nostrils flared with each breath. He breathed in quick, shallow drafts, his mouth open to gulp each bit of air. About the mouth, a pallor. The hollows there were whitened. The mouth was without the least moisture, the teeth coated with a paste of brown and dirty saliva. Inside, the tongue had wilted. N. removed the boy's clothing and exposed the shiny, mounding abdomen. How ashamed it looked! The light pressure of N.'s palm on the abdomen aroused the boy. The roving eyes slowly found their focus, and he saw the man standing over him. All at once the thin little snout crinkled in beggary, and with each quick breath came the mewing of a cat. The abdomen was hot and tight.

Mona watched him standing over the body of the child, watched his fingers playing the belly lightly, counting, receiving. To her, his fingers were like the growing tips of a plant reaching into the unknown. There was something prophetic about those fingers, willing the events here to happen. Now and then she saw him look away into the distance, as though he had left the room, descended into a furrow on his own brow. And all the while his hands never left the body of the young boy. To do so would be to surrender himself to the uprooting of a fierce wind. Only the clean heat of the boy's flesh kept him from being whipped off into a wild air.

"Make ready," he spoke over his shoulder.

The boy was weightless in his arms, a wraith. He laid him upon the table and straightened the thin legs, slid a pillow beneath the head. He might have been arranging a corpse. Taking soap and water, N. bathed the child's belly with solicitude, taking care not to press. The child's small penis stirred in answer to some uninterpretable memory. The veins in the child's skin were green. For a long time he washed the child, repeating the same movements over and over, rubbing in a circular pattern, beginning at the navel and ever increasing the size of the circle until the whole belly was covered. N. could feel the sweat blistering his forehead. He was trembling. I cannot do it, he thought, as though the announcement of his imposture had robbed him of the way to heal this boy. Better for him to die without me as his executioner. It would be another murder. Just then N. felt the handle of a knife slip into his hand. Had it leaped there? His fingers closed around the instrument, each one falling comfortably into rank.

N. laid the belly of the knife against the belly of the child between that pouting navel and the flare of the right hip. He pressed and drew, feeling the exquisite sharpness of the cut in his own flank. Again and again he sliced, each time descending deeper into the body of the child. He knew the landmarks—fat, muscle and the last, the glistening peritoneum. What he sought would be just beneath this.

Mona soaked up the blood with her cloths, and held the incision apart for him to see. Another pass with the knife, and . . . pus. Up and out it billowed and rolled. N.'s hands were covered with it. It ran across the boy's unbearded pubis to fill his groin. It slipped from the sides of the table to the floor. The boy's belly seemed to deflate before his eyes. The breathing eased. The pulse at the groin was rapid and strong. N. propped open the wound with long wicks of waxed cloth, and stepped back from the table. It was done. Mona covered the wound with cloths and tied it around. When she had finished, she turned to see that N. had slumped into a chair. His whole body shook, as though he had become infected with the pus from the child's abdomen. The woman saw that the eyes of the man were caged foxes prowling in slatted light. And she was afraid. She took a pitcher and started out for the square. She knew that she must bring back some water from the trough. She ran all the way, filled her pitcher, and raced back to the cottage. The door was open. The child lay sleeping peacefully on the bed. N. was gone. The empty chair told her that he would never come back.

N. turned abruptly from the path and stepped into the darkness. He felt the forest enter him the way night enters a cottage, through the window, the door, the chimney. Or the way the sea enters a sinking ship. He was absolutely calm, taking upon himself the patience and renunciation of the trees. Out there, in the village, he had given what he had, what had, on that panting, rabid day three years before, been given to him. He remembered Mona flogging him with her howls, himself struggling to create her with his tongue. He had no more of it left. The surgery he had just performed on the boy was the last of it, the final act of incision and drainage. Somehow he had always known that it would be here in the forest. As a child he had been both terrified and charmed by the tracklessness of it. Every blade and twig wanted him. The leaves waited to lick him clean. From far away he heard the sound of men shouting to each other. There was the barking of dogs. Once again, he was a fugitive. He crashed ahead through the underbrush, weaving among the trees. The muddy, reluctant floor gave back each of his footsteps. From this mud his boots sucked up phantoms. No sooner were these freed from the wet earth than they, too, joined in his pursuit. On and on he slogged. He would not choose the exact place; he would know it when he arrived. He had only, when the moment came, to turn and face it without imposture. At the end point of life was honesty. The low branch of a beech tree swept across his chest, pressing him gently to the ground. There, caught and flickering in a thicket of his own planting, N. turned and lay on his back, letting his fingers enjoy the shallow mud. High above, specks of blue sky, fragments of another world from which he was already receding.

He remembered the cobblestones of the square dancing about his feet. He remembered leaning over the trough, dipping out the secrets of the river and listening to the heavy drag of water far below. It had the solidity of ice. He remembered it streaming through his body until he himself was transparent and permeable, like a fish. He had felt then that he was in the process of *becoming* —exactly what he did not know, something between the golden sun and the dark river, something like the moon, a midmost thing that hides itself in the light of the day. Now N. turned to lie face down, pressing his cheek into the wetness. The cool mud was a healing poultice. He had never known such a sense of well-being, so exultant a physical health. All at once he became aware of a sound at his ear, repeating and repeating. It had exactly the soft, muffled click of old Rolf's dislocated shoulder as it snapped back into place. He held himself to that sound for a long minute as though it were a plank that had been cast to him in a raging sea, then relinquished it with all the rest. N. smiled. Who would have thought it would be old Rolf who would make the sound that would usher him away. And make it not with his mouth, but with the capsule and cartilage of his shoulder? It seemed to him a sound for which he had been listening all his life, and that had always eluded him, but that now at last he had heard.

Now N. turned his head and pressed his face into the floor of the forest. Deeply he inhaled, drawing the mud in, feeling it slide at the back of his throat. There was no pain, no choking, as though water were passing over the gills of a fish. Nothing but immense and endless comfort, a sense of being reclaimed. No pain! But this was the death of a righteous man. He had not earned it! Pushing in, he filled his lungs with the benevolence.

For three days they searched, crashing through the woods, following each flutter and wave, sniffing every wind, scanning the sky for smoke, bounding up toward some movement and finding only each other. The dogs never caught his scent. As well hound the moon across the sky, the people said later. At last they gave up. They left him to the forest.

From the day of his disappearance, the stories began. That day, it was said, all the pigeons of the square twined and wove through the streets of the village, then flew away. Nor had they returned. A town without pigeons is a dead town, the old women said, and consoled themselves with weeping together. The villagers whom N. had treated were envied and deferred to. Whenever people gathered, someone was persuaded (without much difficulty) to tell once more the story of his healing. Come on, Rolf. It's your turn. Tell how he put your arm back in its socket. And Rolf would step forward a bit and tell again how N. had taken off his boot and his sock, how he himself had waited as a child waits to be tickled until he felt (ah!) that heel snuggling into his armpit. The pressure of N.'s bare flesh against the hollow of his own was snug and more perfect than anything his old body had ever known; then there was the pulling, until the little click at the end. And as he remembered, Rolf's left hand would rise to the place where N.'s heel had rested, as though to recapture the touch.

A woman told how he had plucked the red sores from her husband's body as though he were gathering flowers. Each one tried to recall the things he had said, his words that should console them through the lives they must now live without him. But they could only remember his deeds; there were no words, nor any writings. Only the remembered pressure of his palm, the rapt penetration of his gaze, the notion that where he had walked, it became brighter.

Wherever he walked, a sweet smell lingered, they said. When he coughed, every door in the village swung open, they said. One man, a hunter, remembered his footprints, and the lingering steam of his breath. The boy, Mihail, the last of them to be touched by him, remembered his own head falling against N.'s chest as he carried him to the table, remembered his own open mouth, and the feel against his cheek of N.'s nipple through the cotton shirt. To another child, who had been healed by dreaming, he was a shape-less vision, with no words or pictures to represent it, yet past all forgetting. And all together they remembered the time that a bear shambled into the village and right into the square. There was the ring of its claws on the frozen stones. The frightened cottage doors had never seen anything like it. And how N. was not afraid, and had gone to it, smiling to see the bear mounding over the trough, dipping its paw into the water, fishing.

What first were recollections became legends. A wind of prophecy blew through the village. There were rumors. Soon N. had entered their dreams, not only the dreams of each one separately, but the dreams of the whole village, in the same guise. No sooner had a woman lain down in her bed than she would see N. bending over the horse trough, or squatting near the road, whittling toward sharpness. Mostly they saw him touching them. It was like a slow contagion of sleep that spread from cottage to cottage until the dreams of every man, woman and child were infested. They reported their dreams to each other. An old woman told of walking along the edge of the forest behind his cottage. She heard a flapping noise as though a blanket were being shaken. And then she had seen him rising in the air to the topmost branches of a tree. Just folded his arms across his chest, like this, she said. And she would show them. And cried out, "Oh!" in a loud clear voice. Then up he went. Oh, yes, it's true, Lubov's cat had gotten stranded. Which is Lubov's? The gray-and-white with six toes. Shut up and let her tell the rest. What does it matter which is Lubov's cat? You want to know, go to Lubov, he'll show you. Come, Granny, tell the rest. And the old woman would fold her arms across her vanished bosom to show how he came down in the same way, holding the cat on his shoulder.

A man lay in the forest with blood pillaring from a wound in his flesh. From his cottage, N. had called over to the column of blood, "Stop!" And it had stopped, frozen. Then he walked to where the man lay and knelt to break off the solid spurt from the clean healed skin.

And this from a young man who had been born a little foolish: he had been setting his traps in the woods just north of the fields when he caught sight of N. standing alone in the middle of a meadow. The youth drew near to greet N., who had not long before stitched up a deep gash in his thigh. But when he drew near, N. was nowhere to be seen. Instead there was a pool of clear water spreading on the meadow where he had been. The water was both transparent and reflecting. Fish and lilies and clouds were contained within it. And the moon. The young man felt a great yearning to enter the pool. He closed his eyes for a moment. When he opened them and gazed again into the pool, the moon was gone, and where it had been there was the pale and luminous face of N. Just as the youth might have entered the water, he saw N.'s head rise from the center of the pool, then his neck and chest, his whole body, and even as he emerged, the pool shrank until it was only the shirt and sleeves the man wore. Then N. turned and saw the young man, and he smiled as though he had just been surprised in an act of play.

Think of it! It had all happened within the space of one breath. Up and out came his head, then his shoulders, arms, legs. Then he drew the water upon himself like a robe with flowing sleeves, until there was no pool at all, only N. standing there smiling, and shaking the last drops from his head.

And Mona, too, lay in her bed, her head filled with the consequences of the kiss that she could never express, when, like a bat, she had fastened on his warm blood and drank him in and in, parting and searching the hair on her lover's chest, becoming absorbed in the mystery of his nipples. She saw him trembling in the grass, all about him shadows lying curled like great beasts. Needles of light pricked his body. And she tasted again his blood in her mouth where she had bitten him.

For her now, the kiss was a thing that existed by itself. Somehow, between his giving of it, and her receiving it, the kiss had sprung into real substance, something made of flesh, that knew her entirely. And Mona remembered and kept secret the body of N. the way she had kept secret his epilepsy, as though she were a priestess whose charge it was to guard a sacred flame. But the others wondered about it and whispered. After all, it was no secret that she had shared his bed. What must it have been like to lie with such a man? And so N. entered the fantasies of the women. They would lie awake in the dark and try to imagine his body, which must have been the most beautiful in the world. They would try to picture long white limbs, not too fleshy or thin, a blend of strength and languor. Oh, what must it have been like? The men too dreamt of it, and their pulses speeded up. Had anyone dared to ask her, Mona would never have said what she had known of his body, known with her eyes, her hands, her teeth even; would not have said that even in his nakedness he had been shielded from her, as though he had been fully clothed; that, even naked, he had seemed to her to wear his own body as raiment.

Mona continued to live in the cottage she had shared with N. For a while the villagers came seeking medicine, but Mona offered none, and soon they stopped coming. As time went by, she withdrew further and further from the life of the village. She left the cottage only to tend her garden, or to get water. On the street, after dusk, she was a shadowy figure with scarves and shawls flapping about her head. It seemed to the others that she had lived in another age, when myths were made, and heroes and gods walked among men. Somehow she had survived, and was here amongst them now, a bit of evidence.

"Ahhhhh! . . . Look!" A woman would see her on the street and she would stand aside to let her pass. Later the woman would send her child to leave a basket of food on the steps of Mona's cottage. But don't knock, she would instruct the child. Just put the food there, and run home. A few of the oldest people remembered her as she had been. They remembered that there had been something wrong with her, and that she had been healed by N. None could quite recall what it had been . . . something about her face . . . but to anyone who caught a glimpse of her, half hidden in scarves, there was nothing to be seen.

In this manner, sixty years went by. Seth had died a year after N.'s disappearance, calling out, it was said, in a voice full of terrible longing. Mona had long since been buried in the garden at the back of N.'s cottage. Some say that she had taken her own life, had thrown herself into the great river on the anniversary of his coming. Early the next year, just before the thaw, a fisherman caught sight of something gleaming just beneath the transparent ice. He chipped away the crust and saw an upturned face. It was Mona, her frozen body intact. But the face was that of a beautiful young woman.

On the anniversary of his coming, the shrine was opened to the public. Visitors came from far and wide. From the capital, even. By carriage and sled and riverboat, they came to dip their cups into the horse trough. Their sips were self-conscious and reluctant. Was the water safe for drinking? The Fountain of Healing, it was called. From there, the pilgrims crossed the well-worn (and often replaced) cobblestones to the village street on which N.'s cottage still stood. Inside, they would ogle the glass cases of polished mouse bones, the strands of hair, braided and waxed, the mortar and pestle, the bed where he had slept, his chair. They were silent, save for the loud clomping of their boots on the bare planking of the floor. The noise irritated Mihail, who would scold them rudely.

"Pick them up, Princess Heavyfoot. Come, come, Twinkletoes. This is a shrine. Show some respect."

Mihail's ill temper was famous. A favorite blasphemy among the pilgrims was to whisper that N. had left some putrid humors inside the old fellow's belly, and that only now were they working their way to the outside. But then they would gaze at the handmade oak table upon which the cranky old man had lain as a child, waiting for N. to lay open his body. Then their eyes would be drawn to the sacristan in awe, and they would forgive him. At precisely that moment Mihail would lift the hem of his shirt, and pull down on his belt to show the scar. The women would draw in their breath, and the men would reach for another coin. It was a good living.

Remains

———————————•◆•———————————

From the flavor of your last let-
ter, I infer that you have attended your first autopsy. It is only
natural that while studying Pathology you dwell upon death. Nor
ought you to be embarrassed to be caught red-handed in an act of
philosophy. Where, if not in a morgue, would a man's thoughts turn
toward his mortality? It will do your patients no harm to have their
doctor feel fraternal about dying. When there is nothing else to do
for a patient, the doctor functions as Charon, mediator between the
living and the dead. He offers to the dying, if nothing else, his
familiarity with death. But you must not peer too much into your
own grave. It is self-indulgent and wasteful. Take your cue from the
men who work in the morgue. They are of a marvelously cheerful
nature. It is always so in every hospital. And, truly, it does no more
good to think of a tumorous tree gnawed by termites than of a tree
in full and shimmering leaf. The autopsy he performs each day, to
the pathologist, is no mere matter of death, but most intensely a
matter of life. Just see how the glow has not wholly relented from

the body on the slab. The grand implication of an autopsy is of the abounding life to come. The dissector in the morgue rummages in this hotel of bones with its corridors and lobbies, and windows curtained with lace, searching for the legacy of the newly dead. He is the deliverer of wisdom.

Lately, I, too, have been dwelling on the end of my days. This is no morbid turn of the mind, I assure you, no fit of melancholia. I do but obey the advice of Montaigne, who insisted that to learn Philosophy is to learn to die. Make no mistake, I am in no way restless for the next world. I like it fine right here. I only wish to lose the sight from my eyes when I have lost the desire to see anything. And had I influence in Heaven, I would bend the ear of God for longevity. Should that request be denied (and say what you will about the power of prayer, sometimes the answer is just plain "No!"), only then would I lower my sights to resignation.

It is all very well for the clergy to intone the certainty that Death is not the end, but just a new beginning, a "pleasant potion of immortality," as Sir Thomas Browne put it. But I myself admit to a bit of skepticism about that. After all, how do they *know*? Frankly, I'd rather spend the next ten years dining on slugs and bitter vetch than the next ten minutes licking my chops in the back of a hearse at the prospect of ambrosia. Besides, I have watched too many of my fellowmen squeeze through that small tunnel whose narrow girth restricts the breath to quick, shallow huffs, suffuses and darkens the lips, glazes the eyes and presses forth groan upon groan from the poor wheezy thing caught in midpassage.

Such is the prettiness that comes to me as I lie in my bed and cannot sleep. Why is it that the airy persiflage of the clergy is so much more palatable during the daylight hours? While at night the smallest allusion to mortality scrapes through the brain like a grappling hook dragging for a drowned man? I am reinforced in my prejudice against matters funerary by having just returned from the still-open grave of a colleague. All during the prayers, stalking from headstone to headstone, and wearing a terrible smirk—the fattest cat I have ever seen.

The whole matter of departure was brought on by an attack of *la grippe,* which explains my tardiness in answering your letter. There is nothing more catalytic toward the birth of a philosophy than *la grippe.* Chills and fever howl all around; splayed against the rocks you hear the beating of terrible wings. Even your dreams are mortuary. In one, I saw the skeleton of a man dangling from the limb of a tree, its bones held together by sun-dried ligaments. A noose looped the cervical vertebrae, and the skull, tilted forward, set its empty sockets upon a red bird (a scarlet tanager?) nesting in its ribs. Unreleased as yet from its wholeness by those kind and seemly vultures devoted to tidiness, the hanged poet sang on, piercing notes that have not been heard since Eden.

With just such apparitions have my dreams been filled. Then, one day, I awakened to a calm, warm breeze. There was the earliest stirring of appetite. Praise be! The beast, distracted by some toothier morsel, had let fall from his jaws my half-eaten carcass. Thus abandoned, I began my recovery. (O winged word!) Now is a time of mingled joy and sadness: Joy, as the storm fades further and further in the distance. Sadness, because . . . well . . . who can forget a glimpse into the mouth of Hell? Try as I might to avoid this remembrance, in unguarded moments my thoughts turn again and again to the old horror. I hear the raven quothing, "Nevermore." And I shudder anew. "Look ahead," my friends say. "Buck up." And I *try* to savor the repair of my words. Honestly, I do. But some of us, like Lot's wife, are given to backward glances.

"Bundle up," my wife had said. "You'll catch your death." And so I almost did. Though *catch* may be the wrong verb here as I have no mind to run after it, even though the absence of a belief in a Hereafter is of some comfort. One can die without the anxiety of the believer, secure in oblivion, ready for Nothing.

Apropos of Lot's wife, I can think of worse fates than to be turned into a pillar of salt. (Sugar would not be acceptable, since it is likely to draw flies.) There is something satisfying to the spirit about being licked away to nothing by the wet, abrasive tongues of deer, or drilled by the wind. For whatever the deer and the wind didn't get, the rain would be the final solution. By the way, I hereby charge you with the disposal of my remains.

Yes, you! Don't look so pale. I am not asking you to hold my sword for me to run upon, though had I the choice of my dispatch, I should ask to die by the sword rather than by the nursing home, as I much prefer a good, clean, lancinating pain to a long listlessness. There is a danger of becoming angry at death as though it were less a friend than an adversary when, in truth, it is neither one more than the other. Say that a man of ninety is brought into the Emergency Room. As he is lifted to a stretcher he makes what is apparently the last of a series of peaceful, slow exhalations. At least there are none to follow it. So long and delicate this final sigh that it is not difficult to imagine the soul being carried out of the man's mouth on the current of it. But we do not permit such a serene exitus. Like guards from whom a prisoner is escaping, we fall upon the miscreant, flailing at his chest, punching, stinging, prodding, electrocuting, demanding that he be recaptured. We forbid him to remain dead. Nor do we have the least inkling of our absurdity, but are caught up in the game of it all. It is a fury to which my heart is no longer equal. I should as soon, I confess it, close the man's eyelids with my thumbs and tiptoe away.

Now to the business: I shall not be dogmatic about the method of disposal. In this, at last, you shall be on your own. If it is to be burial, I reject embalming, and would prefer to remain unboxed. Take care that the earth be packed snugly all around, an even pressure that will afford the coziness I love, and would also prevent any slippage away. I should not like to face eternity with a piece missing. You know how I am about cozy. Give me a chimbly cottage with a feather of smoke and, outside, all around, snow on the prowl, sniffing at the windows.

Justest, I suppose, for an unregenerate smoker, would be cremation. To be inhaled by God—his tobacco. After the ignition, you could watch from a neighboring ridge while "me" wafts toward the pines, spreading out thin before going over to the other side. At the last, only a bit of ash for you to blow away.

But if it is to be interment, I shall allow you a bit of wit. Secure, if you like, one of my long bones—the left femur, I think—and after a sensible period of time, whittle from it a sort of flute with finger holes and all, and lacquer it to a fare-thee-well. What fun to think of Rimski-Korsakov blown out of what once occupied my pants. Yeats wanted to be a mechanical bird hammered out of gold, so he could be immortal. Me, I'd rather be of my own stuff. Then with a bit of breath and some finger work, I'd sing on and on. Flesh liquefies into the occasional tears of the bereaved, while bone dawdles away eternity, powdering off a grain at a time. In the durability of his workstuff, the orthopedist owns the longest calling.

The more I think about it, the better I like burial in the ground. People are of two kinds, you know. Those that love nothing more than to be wrapped up snug by their environs, and those who yearn to be unfettered, airborne, wafted. Knowing this, it is not at all amazing that some prefer interment while others opt for cremation. It is not a reasoned choice; it is a matter of temperament.

You agree? Then burial it is. Let the worms sign my ribs, and the pretty foot of a lizard imprint my mandible, fossils for some as yet unborn anthropologist. These creatures—ants, lizards, worms—do but ransack the corpse in hopes of finding what is hidden there—a jewel or a soul or some bit of evidence that this man has loved. Nothing lasts as it is anyway, but advances toward hard stability from which it can be retrieved one day by curious and wondering fingers. An anthropologist, sifting through the sands of Arabia, comes upon King Solomon's finger bone. He slips a ring from the dead white phalanx, slides it onto his own finger and all at once there is the faint enduring scent of frankincense. It has all returned. The found object is lit by memory. It lives again.

Intellectually, I admire the Zoroastrians, whose method of disposal is to serve up their dead alfresco to the vultures that line the rooftops waiting for just such a feast. It seems both tidy and sensible. These vultures are truly in the service of man. The recycling of animal matter is economical. But where, I wonder, are the creatures who dine on the vultures in their time? Are there any? Lice, I suppose. Heaven, in its wisdom, made the flesh of the buzzard unpalatable to all but the most ravenous of jackals, thus assuring sufficient attendance at Zoroastrian banquets of these sweeps and chars.

Not a pretty sight, you say? I have seen worse, most having to do with the deprivation of love. Still, my admiration for this Asiatic practicality is tempered by Occidental sentiment. I confess to a love of graveyards. There is nothing so romantic as a rose garden where the bodies of young soldiers are buried. The roses there are celestial. They glow as no others do. You gaze at them, inhale their fragrance . . . and you remember old faces and old deeds. It is the next thing to reincarnation.

Now, as to the matter of my own autopsy. Hmmm . . . I have thought of it . . . and I prefer not to.

"Hypocrite!" you cry. "You, above all others—you should see the need!"

"Yes, but there is a greater need. To myself."

"But why not?" you ask. You are indignant, outraged.

"With homage to Bartleby the Scrivener," I answer, "I just prefer not to."

All this dwelling on remains! By the time Death comes to claim me, I shall already have been dead for many years—from long immersion in the habit of it.

ABOUT THE AUTHOR

Richard Selzer was for many years a surgeon practicing in New Haven, Connecticut, where he was also on the faculty of the Yale School of Medicine. He was born in 1928 in Troy, New York, and is a graduate of Union College, Albany Medical College, and the Surgical Training Program of Yale University.

Dr. Selzer's other books include *Down from Troy, Imagine a Woman and Other Tales*, and *Raising the Dead*. He lives with his wife in New Haven.